Karsten Kuritz

Analysis and Control of Cellular Ensembles

Exploiting dimensionality reduction in single-cell data and models

Logos Verlag Berlin

Bibliographic information published by the Deutsche Nationalbibliothek

The Deutsche Nationalbibliothek lists this publication in the Deutsche Nationalbibliografie; detailed bibliographic data are available on the Internet at http://dnb.d-nb.de

D93

ISBN 978-3-8325-5209-1

Logos Verlag Berlin GmbH
Georg-Knorr-Str. 4, Geb. 10,
D-12681 Berlin
Germany

Tel.: +49 (0)30 / 42 85 10 90
Fax: +49 (0)30 / 42 85 10 92
http://www.logos-verlag.de

Analysis and Control of Cellular Ensembles

Exploiting dimensionality reduction in single-cell data and models

Von der Fakultät Konstruktions-, Produktions- und Fahrzeugtechnik
und dem Stuttgart Centre for Simulation Technology
der Universität Stuttgart zur Erlangung der Würde eines
Doktor-Ingenieurs (Dr.-Ing.) genehmigte Abhandlung

Vorgelegt von

Karsten Kuritz

aus Mutlangen

Hauptberichter:	Prof. Dr.-Ing. Frank Allgöwer
Mitberichter:	Prof. Dr.-Ing. Jan Hasenauer
	Prof. Dr.-Ing. Julio Saez-Rodriguez

Tag der mündlichen Prüfung: 4. September 2020

Institut für Systemtheorie und Regelungstechnik
der Universität Stuttgart

2020

To Verena
who celebrated the successful completion
of my PhD more than I did.

Acknowledgments

There are many people who supported me during my time as a PhD student at the Institute for Systems Theory and Automatic Control (IST) and to whome I would like to express my gratitude.

First and foremost, I want to express my gratitude to Prof. Frank Allgöwer for his guidance and support throughout the years, as well as for granting extraordinary scientific freedom. At the IST I found a stimulating and friendly environment with brilliant and at the same time open-minded visitors and colleagues. My colleagues supported my journey over the years and I am thankful that I could be a part of this very special group of people. Therin, special thanks goes to my long time office mate Simon Michalowsky, as well as to Rainer Blind, Hans-Bernd Dürr, Gregor Goebel, Wolfgang Halter, Dirke Imig, Steffen Linsenmayer, Max Montenbruck, Anne Romer, Daniella Schittler, Caterina Thomaseth, Philipp Wenzelburger and Shen Zeng for the great time we had togother, the great discussions and fruitful collaborations.

I went through many ups and downs of my scientific career together with my experimental partners. Hence, I am thankful for long-term trust and support from my collaborators at the Institute of Cell Biology and Immunology, Nadine Pollak, Daniela Stöhr and Prof. Markus Morrison. Furthermore, I want to express my gratitude to Prof. Hana El-Samad who hosted my reasearch visit at the University of California, San Francisco. My visit became a wonderful experience in terms of both my friendship and my research with Alain Bonny and João Fonseca.

Numerous other people such as friends, students and scientists contributed in different ways to my PhD. I am grateful for their companion and sorry that I cannot name all of them here.

I also want to wholeheartedly thank my parents Aki and Werner as well as my siblings Anthea and Arvid for all their love and support during my life. Finally, I want to express my deepest gratitude to Verena who supported my throughout the years. Verena and our children Tadeus and Gwendoline substantiated the reasoning and meaning of my work and thus contributed significantly to this success.

Stuttgart, October 2020
Karsten Kuritz

Table of Contents

Abstract ix

Deutsche Kurzfassung xi

1 Introduction 1
- 1.1 Motivation and focus . . . 1
- 1.2 Contributions and Outline . . . 3

Part I Analysis

2 Reconstructing temporal and spatial dynamics from single-cell pseudotime 11
- 2.1 Background and problem formulation . . . 12
- 2.2 MAPiT - *MAP* of pseudotime *i*nto real-*T*ime . . . 12
- 2.3 Reconstructing spheroids with MAPiT . . . 15
- 2.4 Analysing cell cycle progression with MAPiT . . . 22
- 2.5 Summary and discussion . . . 25

3 Cell cycle analysis with ergodic principles and age-structured population models 29
- 3.1 Background and problem formulation . . . 29
- 3.2 Age-structured population models . . . 30
- 3.3 Relationship between ergodic analysis and age-structure population models . 31
- 3.4 Cell cycle-dependent expression of Cyclin B1 . . . 37
- 3.5 Summary and discussion . . . 42

4 Cell cycle progression inference 45
- 4.1 Background and problem formulation . . . 45
- 4.2 Cell cycle progression inference with sensitivities . . . 47
- 4.3 Evaluation with in silico data . . . 51
- 4.4 Analysis of cell cycle progression . . . 53
- 4.5 Summary and discussion . . . 54

Part II Control

5 Passivity-based ensemble control for cell cycle synchronization 59
- 5.1 Background and problem formulation . . . 59
- 5.2 Theoretical foundation . . . 60
- 5.3 Ensemble control for cell cycle synchronization . . . 62
- 5.4 Computational studies . . . 66
- 5.5 Summary and discussion . . . 71

6 **Ensemble control for cellular oscillators** 73
6.1 Background and problem formulation 73
6.2 Ensemble control for oscillator moments 76
6.3 Computational studies 80
6.4 Summary and discussion 83

7 **Conclusions** 85
7.1 Summary and discussion 85
7.2 Outlook 88

Appendices 91

A **Technical background** 93
A.1 Dynamical systems and control theory 93
A.2 Weakly connected oscillators 95
A.3 Fourier coefficients and circular moments 98

B **Experimental protocols and data preprocessing** 101
B.1 Ordering cells in pseudotime 101
B.2 Establishment of NCI-H460/geminin cell line 103
B.3 Live cell microscopy 103
B.4 Flow cytometric analysis 104
B.5 Construction of the average cell cycle trajectory 104
B.6 Estimation of constant speed and noise from cell cycle lengths 105

C **Spheroid growth model** 107

D **Cell cycle model** 111

E **Technical computations and proofs** 113
E.1 Speed equation 113
E.2 Derivation of ODE for $\varphi : s \mapsto x$ 114
E.3 Calculation of division rate from cell cycle length distribution 115
E.4 Proof of Lemma 6.1 116

Bibliography 117

Publications of the Author 127

Abstract

This thesis is dedicated to the analysis and control of cellular ensembles. A collection of nearly identical copies of individual cells is called a cell ensemble. In an ensemble system the population may only be manipulated or observed as a whole. This description is suitable for single-cell experiments where measurement data consists mostly of snapshots which are sought as representative sample of the population. Furthermore, it is often inherent to the experimental setups or treatment scenarios, that cells within the population cannot be manipulated individually, but only through a common signal, for example, by a common stimulus through a drug treatment. In this thesis we approach the problem of analysing and controlling cellular ensembles by considering 1-dimensional dynamics of biological processes in high-dimensional single-cell data or models.

The first part of this thesis addresses the quest for real-time analysis of biological processes within single-cell data. While methods exist to describe the progression and order of cellular processes from snapshots of heterogeneous cell populations, these descriptions are mainly restricted to arbitrary pseudotime scales. We therefore developed a universal transformation method that recovers real-time dynamics of cellular processes from pseudotime scales by utilising knowledge of the distributions on the real scales. We exploit this concept in depth for cell cycle studies of unsynchronised cell populations. Furthermore, we extend the results towards learning of changes in cell cycle progression in response to treatments. Our universal tools recover temporal or spatial cellular trajectories in high-dimensional single-cell experiments, thereby enabling analysis of process dynamics in high-throughput and high-content setups.

In the second part we present novel ensemble control algorithms for populations of cellular oscillators. Many diseases including cancer and Parkinson's disease are caused by loss or malfunction of regulatory mechanisms in an oscillatory system. Successful treatment of these diseases might involve recovering the healthy behavior of the oscillators in the system, for example, achieving synchrony or a desired distribution of the oscillators on their periodic orbit. With single-cell data as input an ensemble control formulation in which a population-level feedback law for achieving a desired distribution is sought. Our systems theoretic approach to this problem results in novel necessary and sufficient conditions for the control of phase distributions in terms of the Fourier coefficients of the phase response curve. Since our treatment is based on a rather universal formulation of phase models, our results are readily applicable to the control of a wide range of oscillating populations, such as circadian clocks, spiking neurons or cells in the cell cycle.

This thesis establishes a connection between the previously separate areas of single cell analysis and ensemble control. Our holistic view opens new perspectives for theoretic concepts in basic research and therapeutic strategies in precision medicine. In the future, we envision the integration of our ensemble analysis and control algorithms into online therapy design schemes to enhance precision medicine and accelerate drug discovery.

Deutsche Kurzfassung

Analyse und Regelung von Zellpopulationen mittels Dimensionsreduktion in Einzelzelldaten und -modellen

In der vorliegende Arbeit werden Methoden zur Untersuchung und gezielten Beeinflussung von Zellpopulationen beschrieben. Konkret werden Zell-Ensembles betrachtet, welche sich aus einer Menge von nahezu identischen Einzelzellen zusammensetzen. Ein Zell-Ensemble kann nur als Ganzes beeinflusst oder beobachtet werden. Einzelzellexperimente, in welchen Momentaufnahmen von repräsentativen Stichproben der gesamten Population untersucht werden fallen in diese Kategorie. Häufig ist es zudem auf Grund der experimentellen Bedingungen nicht möglich gezielt einzelne Zellen zu beeinflussen. Es ist vielmehr so, dass externe Signale, wie beispielsweise medikamentöse Behandlungen in gleicher Weise auf alle Zellen einer Population wirken. In dieser Arbeit behandeln wir Fragestellungen zur Analyse und gezielten Beeinflussung von Zell-Ensembles, indem biologische Prozesse auf 1-Dimensionalen Mannigfaltigkeit in hochdimensionalen Einzelzelldaten oder -modellen beschrieben werden.

Der erste Teil dieser Arbeit befasst sich mit der Echtzeitanalyse von biologischen Prozessen aus Einzelzelldaten. Eine Vielzahl von Analysemethoden ist in der Lage die Abfolge zellulärer Prozesse aus Einzelzelldaten abzuleiten. Eine Schwäche dieser Methoden ist es, dass die einzelnen Zellen der untersuchten Population auf einer nahezu beliebigen Pseudo-Zeit Skala geordnet werden. Die Zustandstransformation die in dieser Arbeit präsentiert wird verwendet Vorwissen über die Verteilung der Population auf dem realen Maßstab um Dynamiken aus Pseudo-Zeit in Echtzeit zu übertragen. Die Anwendung dieses Konzepts auf Zellpopulationen in der exponentiellen Wachstumsphase ermöglicht die Echtzeitanalyse von Zellzyklusdynamiken aus Momentaufnahmen von Zellpopulationen in ihrer Gleichgewichtsverteilung. Durch die Beschreibung der Zellpopulationen mittels Altersstrukturmodellen kann darüber hinaus eine behandlungsbedingte Veränderungen der Zellzyklusdynamik erfasst werden. Aufgrund der allgemeinen Gültigkeit der vorgestellten Methoden lassen sich zelluläre Prozesse neben der zeitlichen auch in ihrer räumlichen Ausdehnung aus Einzelzelldaten rekonstruieren. Dies ermöglicht die Analyse hochdimensionaler dynamischer Prozesse in automatisierbaren Hochdurchsatzverfahren.

Im zweiten Teil werden neuartige Regelgesetze für die gezielte Beeinflussung oszillierenden Zellpopulationen präsentiert. Viele Krankheiten, darunter Krebs, Parkinson und Herzerkrankungen werden durch den Verlust oder die Fehlfunktion von Regulationsmechanismen in Systemen mit periodischen Schwingungen verursacht. Eine erfolgreiche Behandlung dieser Krankheiten könnte durch die Wiederherstellung des gesunden Schwingungsverhalten der Oszillatoren erreicht werden, wobei die Zellen der Population beispielsweise synchronisiert, oder wunschgemäß auf ihrer Umlaufbahn verteilt werden. Hierfür wird ein Ensemble-Regelgesetz entwickelt, welches ausgehend von Einzelzelldaten ein populationsweites Eingangssignal erzeugt. Notwendige und hinreichende Bedingungen für das Erreichen einer gewünschten Verteilung ergeben sich dabei aus den Fourier Koeffizienten der phasenabhängigen Signalantwort. Die Ergebnisse der Arbeit sowie die daraus resultierenden Regelgesetze sind auf Grund der allgemeinen Gültigkeit des verwendeten Phasenmodells anwendbar auf ein breites Spektrum oszillierender Zellpopulationen, beispielsweise die innere Uhr, neuronale Aktivitätsmuster und Zellzyklusdynamiken.

Zukünftig, könnte die Kombination der beschriebenen Analyse- und Regelmethoden in personalisierte Echtzeittherapiepläne eingehen und somit Bestrebungen hin zu einer umfassenden Präzisionsmedizin fördern, sowie die Entwicklung neuer Medikamente durch die erlangten Erkenntnisse beschleunigen.

Chapter 1

Introduction

1.1 Motivation and focus

Biological research, and natural sciences in general, commonly aquire new knowledge by testing a hypothesis against experiment data. A multitude of experimental methods and protocols has been developed over time to reveal the composition of biological systems. Analysing complex biological systems aiming at a deeper understanding of the processes in living systems requires the integration of experimental and computational research (Kitano 2002). Particularly rich information is present in high-dimensional single-cell data. Such data is generated by methods like microscopy, flow cytometry or single-cell RNA sequencing, where the abundance of up to thousands of cellular components for every individual cell in a population is measured. These experiments thus capture the heterogeneity present in a cell population.

When talking about reasons for heterogeneity in biological systems, one often differentiates between intrinsic and extrinsic noise. Intrinsic noise is characterized by the absence of, or only short time correlation between quantities in identical cells of a population. It is thought of as a system inherent property that emerges from stochastic fluctuations in biochemical reactions involving low copy number of genes and thereby causing heterogeneity in a population. Extrinsic noise on the other hand exhibits long time correlations of the quantities in a population (Elowitz et al. 2002; Swain et al. 2002; Munsky et al. 2009; Iversen et al. 2014). Such persistent variance between cells in a population is caused by various factors including the local environment or the history of cells (Snijder et al. 2009; Gut et al. 2015; Sandler et al. 2015). Examples for extrinsic noise caused by history of cells are the cell cycle or cell differentiation processes. Therein, the cellular components vary depending on the progress of individual cells along the cell cycle or differentiation pathway.

Experimental single-cell data that is randomly spread around the population average is observed for a stationary population with prevailing intrinsic noise (Figure 1.1 a). In contrast, single-cell data may be spread around a path in the data space, if cells in the population are additionally at different stages of a process (Figure 1.1 b).

Hence, information about a biological process is present in single-cell data where the population is spread over that process in terms of: (1) shape of the path in data space, (2) the distribution of cells along the path and (3) variance of the population around the path. Furthermore, by changing these characteristics a cell population can be manipulated to achieve a desired behavior. This may for example be inhibition of cell growth in cancer treatment, neuron synchronization in jet-lag, or neuron desynchronization in Parkinson patients.

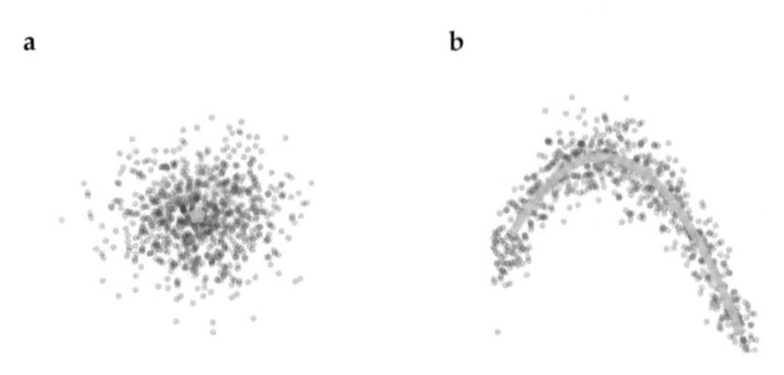

Figure 1.1. Distinction between heterogeneity in single-cell data originating from (**a**) random noise in a stationary cell population or (**b**) an additional underlying process.

Systems and computational biology exploit mathematical models to understand and predict the dynamics of biological systems. A common way to mathematically model molecular processes in a cell is via ordinary differential equations (ODE) models (Klipp et al. 2009; MacArthur et al. 2009). These models describe the concentration change of cellular components in a single cell, or the average cell of a population under deterministic dynamics

$$\begin{aligned} \dot{x}(t) &= f(x(t), u(t)) \,, \\ x(0) &= x_0 \,. \end{aligned} \tag{1.1}$$

Therein, the state variables $x(t) \in \mathbb{R}^n$ represent different molecular species in the cell which can be affected by external inputs $u(t) \in \mathbb{R}^l$ such as media, drugs, optogenetic cues or environmental factors. The dynamics are determined by the vector field $f\colon \mathbb{R}^n \times \mathbb{R}^l \to \mathbb{R}^n$.

A collection of nearly identical cells, also termed an ensemble, may be modeled as multi-agent system, with each agent being a dynamical system with dynamics given by Eq. (1.1). Mathematically, an ensemble can also be described in terms of a density function over a state space $p(x,t)$ (Wiener 1938) as shown in Figure 1.2. The dynamics are governed by partial differential equations, belonging to the class of *Liouville equations* (Gyllenberg and Webb 1990; Brockett 2012) of the general form

$$\begin{aligned} \partial_t p(x,t) &= -\langle \partial_x, f(x,u(t))\, p(x,t) \rangle \,, \\ p(x,0) &= p_0(x) \,, \end{aligned} \tag{1.2}$$

equipped with boundary conditions. The transport equation Eq. (1.2) describes how a density $p_0\colon \mathbb{R}^n \to \mathbb{R}_{\geq 0}$ of initial states is advected with the flow of a nonlinear differential equation of the form $\dot{x}(t) = f(x(t), u(t))$.

Brockett (2012) defines an ensemble system as a collection of nearly identical dynamical systems which admit a certain degree of heterogeneity, and which are subject to the restriction that they may only be manipulated or observed as a whole. This description is suitable for single-cell experiments where measurement data consists mostly of population snapshots which are sought as representative sample of the population. A population snapshot, taken at a given instance in time, provides a vast number of output measurements. Yet, at the same time information relating a measurement to the individual system that produced the measurement is not provided, since the measurement process usually results in killing the cell, making it impossible to measure that cell again. Furthermore, it is often inherent to the experimental setups or treatment scenarios that cells within the population cannot be

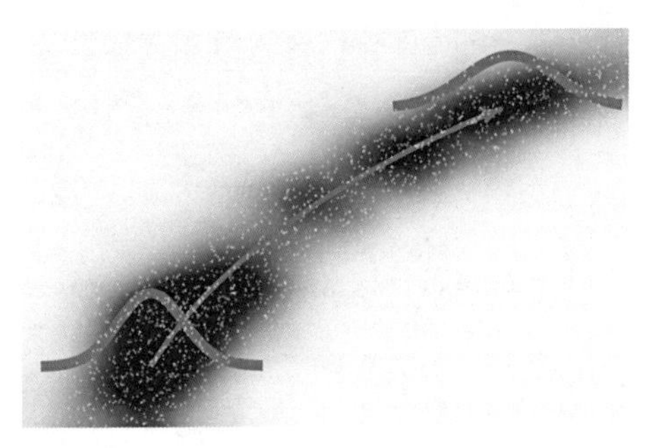

Figure 1.2. Flow of single cells and representation of cellular ensembles.

manipulated individually, but only through a common signal, such as a common stimulus through a drug treatment. Control theory employs mathematical models to derive feedback controllers with the goal to achieve a desired system behavior. The above limitations then lead to the ensemble control problem where we derive a broadcast input signal based on snapshot data to control a heterogeneous cell population.

Major goals of biological and medical research are to (1) understand the dynamics of biological processes, which means to determine the vector field $f(x(t), u(t))$, and (2) control the dynamics, which means to steer the state of a single cell $x(t)$ or a cell population $p(x, t)$ to a desired behavior. This thesis presents analysis and control methods based on single-cell data for cellular processes in heterogeneous populations. We will introduce the underlying concepts in cell cycle studies. In particular, we (1) present a theory to identify the local vector field along a biological process observed in single-cell data and (2) derive an ensemble control formulation to achieve any desired distribution in a population of cellular oscillators.

By reducing the dimensionality of the original data or model to a 1-dimensional manifold, these tasks boil down to problems where we want to analyse or control the distribution of cells along a given process in 1-D. Our results obtained in a 1-dimensional framework are transferable to higher dimensions by a homeomorphism between the description of a process in 1-D and the corresponding high-dimensional model or data.

1.2 Contributions and Outline

In this section we present the outline and summarize the main results and contributions of the individual chapters (illustrated in Figure 1.3). The thesis is structured in two parts: Part I compromising Chapters 2 to 4 contains results on the analysis of biological processes with single-cell data. Part II compromising Chapters 5 and 6 picks up these results for the development of ensemble control algorithms for oscillating cell populations.

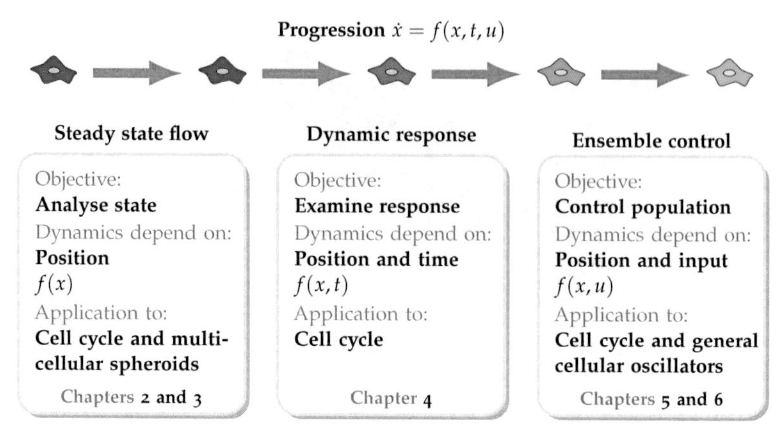

Figure 1.3. Overview of the thesis

Part I Analysis

Chapter 2: Reconstructing temporal and spatial dynamics from single-cell pseudotime

Chapter 2 presents the fundamental theoretic concept to perform real-time analysis with single-cell snapshot data of heterogeneous cell populations spread over different stages of a biological process. First, we introduce the pseudotime representation of single-cell data and briefly discuss its properties and limitations. We then address the arbitrariness of the pseudotime scale by introducing the measure-preserving *map* of pseudotime *i*nto real-*t*ime, in short *MAPiT*. After discussing properties of the method we apply MAPiT for a temporal scale in cell cycle studies and for a spatial scale representing the distance from the surface in cell spheroids. This chapter is based on the publications Kuritz et al. (2017) and Kuritz et al. (2020b). Our main contributions in this chapter are the following:

- MAPiT provides a theoretic basis for the relation of pseudotime values to real temporal and spatial scales.
- MAPiT recovers the progression rate on the process manifold in snapshot data from heterogeneous cell populations.
- By applying MAPiT on two completely distinct problems we demonstrate its universal nature and broad applicability for the analysis of cellular processes.

Chapter 3: Cell cycle analysis with ergodic principles and age-structured population models

Chapter 3 examines the results on cell cycle analysis from Chapter 2 in a dynamical systems perspective. Therein, the evolution of the distribution in pseudotime stems from the desciption of progression of a single cell through its cell cycle by stochastic differential equation (SDE). Based on ergodic theory, we derive a transformation of this model to age-structured population models. We do this for scenarios without noise, with extrinsic noise and with intrinsic noise in the description of cell cycle progression. The scenario without noise recapitulates the results from Chapter 2. For the scenarios with noise we derive methods to infer noise strength and incorporate this information in the transformation. Finally, we discuss the different approaches and evaluate the results against live-cell microscopy data. This chapter is based on the publication Kuritz et al. (2017). Our main contributions in this chapter are the following:

- We establish the relation between age-structured population models and cell cycle analysis with snapshot data.
- We derive inference algorithms for intrinsic and extrinsic noise in cell cycle progression.
- We present a transformation from pseudotime to real-time by convolution of the distributions which takes progression noise into account.

Chapter 4: Cell cycle progression inference

Chapter 4 presents an extension of MAPiT to non-stationary processes. The chapter deals with the specific example where we want to infer altered cell cycle progression in response to treatments. We first motivate the problem and describe data processing steps, which are based on MAPiT. Next, we formally describe the partial differential equation (PDE) model and the estimation problem for the inference of a time- and position-dependent cell cycle progression rate. We present a way to efficiently solve the optimization problem by providing parameter sensitivities. Finally we discuss properties of our method and demonstrate its capability with one artificial and two experimental data sets. This chapter is based on the publications Kuritz et al. (2020a). Our main contributions in this chapter are the following:

- We present a computational framework that allows the inference of changes in cell cycle progression from static single-cell measurements.
- We efficient solve the estimation problem for the time- and cell cycle position-dependent progression rate by caclulating parameter sensitivities.

Part II Control

Chapter 5: Passivity-based ensemble control for cell cycle synchronization

Chapter 5 introduces an ensemble control algorithm for cell cycle synchronization. First, we introduce the research topic and formulate the problem in terms of the reduced phase model approach and its relation to MAPiT. Next, we derive the passivity-based control algorithm and provide necessary and sufficient controllability conditions for cell cycle synchronization. Finally, we evaluate the approach in a realistic individual-based simulation framework where we observe parameter ranges in which synchrony is achieved despite the naturally occurring heterogeneity. This chapter is based on the publications Kuritz et al. (2018a) and Kuritz et al. (2018b). Our main contributions in this chapter are the following:

- We introduce a state transformation for age-structured population models to enable passivity-based controller design.

- We derive an ensemble control algorithm to achieve cell cycle synchronization with broadcast input signals.
- We present a theoretic condition for controllability and practical parameters ranges for the synchronization in realistic setups.

Chapter 6: Ensemble control for cellular oscillators

Chapter 6 presents control strategies for the manipulation of processes in heterogeneous cell populations, in particular, cellular oscillators. We introduce a population-level feedback that is capable to achieve any desired distribution of cellular oscillators on their periodic orbit. First, we motivate the research topic including a summary of major results in the field of ensemble control. After deriving the control algorithm we provide controllability conditions which we discuss for some real-world systems. Finally, we present the performance and limitations of the algorithm in computational studies. This chapter is based on the publications Kuritz et al. (2018a) and Kuritz et al. (2019). Our main contributions in this chapter are the following:

- We present an ensemble controller to achieve any distribution of cellular oscillators on their limit cycle.
- We derive controllability conditions for convergence based on properties of the phase response curve.
- Our controller is applicable to many problems, such as, phase shifting of the circadian clock, cell cycle synchronization or desynchronizing of spiking neurons in Parkinson's disease.

Chapter 7: Conclusions

Chapter 7 summarizes the main results of this thesis, presents the conclusions and indicates possible directions for future research.

Appendices

The results in this thesis build up on various theoretic concepts. We describe these concepts in detail in Appendix A. Therein, we briefly introduce the system theoretic basis and control theoretic concepts in Appendix A.1, the concept of reduced phase models in Appendix A.2 and Fourier analysis and circular moments for circular data in Appendix A.3. Furthermore, we cover experimental protocols and data processing procedures in Appendix B, including a review of the basic concepts of trajectory inference algorithms and the resulting pseudotemporal ordering in Appendix B.1. Appendices C and D provide a summary of the spheroid growth model and the cell cycle model, respectively. Finally, Appendix E compromises technical computations and proofs which we do not present in the main chapters in order to improve readability.

The chapters in this thesis are based on severel publications. These publications were addressed to different audiences. For example, Chapter 2 is based on Kuritz et al. (2020b) which had a diverse but application oriented audience in mind. On the other hand, Chapter 6 is based on Kuritz et al. (2019) which is part of a special issue on *Control and Network Theory for Biological Systems* and thus aimed for a more theory oriented audience. Likewise, we addressed the different chapters in this thesis to different audiences depending on our understanding of the main contributions and their impact on the respective community.

Part I

Analysis

Chapter 2

Reconstructing temporal and spatial dynamics from single-cell pseudotime

This chapter is based on the publication:

> Karsten Kuritz et al. (2020b). 'Reconstructing temporal and spatial dynamics from single-cell pseudotime using prior knowledge of real scale cell densities'. In: *Scientific Reports* 10.1, p. 3619. DOI: 10.1038/s41598-020-60400-z.

Modern cytometry methods allow collecting complex, multi-dimensional data sets from heterogeneous cell populations at single-cell resolution. While methods exist to describe the progression and order of cellular processes from snapshots of such populations, these descriptions are limited to arbitrary pseudotime scales. Deducing real-time dynamics from pseudotemporal ordering however is challenging owing to the arbitrariness of the pseudotime scale. In this chapter, we introduce the measure-preserving *map* of pseudotime *i*nto real-*t*ime, in short *MAPiT*. MAPiT provides a universal transformation method that recovers real-time dynamics of cellular processes from pseudotime scales by utilising knowledge of the distributions on the real scales. As use cases, we applied MAPiT to two prominent problems in the flow-cytometric analysis of heterogeneous cell populations: (1) recovering the spatial arrangement of cells within multi-cellular spheroids prior to spheroid dissociation for cytometric analysis, and (2) recovering the kinetics of cell cycle progression in unsynchronised and thus unperturbed cell populations. Multicellular spheroids grown from cancer cells are widely used as avascular tumour models and proved to be a valuable experimental system, closing the gap between *in vitro* and *in vivo* studies. However, cumbersome preparation of spheroid slices for imaging experiments with limited availability of fluorescent probes restricts the practicability of spheroid experiments. By recovering the spatial position MAPiT reverts the loss of spatial information in single-cell experiments. This enables high-throughput and high-content studies of 3-D-spheroid models. Since MAPiT provides a theoretic basis for the relation of pseudotime values to real temporal and spatial scales, it can be used broadly in the analysis of cellular processes with snapshot data from heterogeneous cell populations.

The experimental data that we present in this chapter was prepared by the Morrison Lab at the Institute of Cell Biology and Immunology at the University of Stuttgart. This chapter is taken in parts from Kuritz et al. (2020b).

2.1 Background and problem formulation

Here, we briefly introduce the pseudotime representation of single cell data and state its main shortcoming which will lead to the problem formulation in this chapter. We provide a comprehensive discussion of the concept of trajectory inference algorithms in Appendix B.1. Single-cell experiments such as flow cytometry, mass cytometry and single-cell RNA-sequencing (scRNA-seq) capture the heterogeneity in cell populations (Klein et al. 2015; Bandura et al. 2009) . The heterogeneity may originate from the fact that the measured cell population is distributed across intermediate cellular states of a biological process, such as cell cycle or cell differentiation (Saelens et al. 2019). This enables the study of biological processes with pseudotime algorithms like CALISTA (Papili Gao et al. 2019), Wanderlust (Bendall et al. 2014), Monocle (Trapnell et al. 2014) or diffusion maps (Haghverdi et al. 2014). These algorithms can capture trajectories in data space by recovering a low-dimensional structure in high-dimensional observations. By ordering cells on a lower dimensional process manifold in the dataspace pseudotime algorithms provide access to the sequence of steps during the process. Common pseudotime algorithms order cells on a pseudotime scale based on a distance metric in the data space, and this metric differs between algorithms (Saelens et al. 2019). Pseudotime values furthermore strongly depend on the measured cellular components. The derived pseudotime thus is a quantitative value of the progression through a biological processes. It is characterized by the relations of high-dimensional observations and in general not equal to the true (time) scale (Weinreb et al. 2018). Hence pseudotime does not directly correspond to real time but is rather a metric in data space of measured cell states. This leads us to the following problem formulation:

Problem 2.1. Given a single-cell snapshot data with a pseudotemporal ordering from a biological process, find a mapping from pseudotime scale to the true (time) scale.

To solve this problem and overcome the arbitrariness of pseudotime scales, we developed *MAPiT* (measure-preserving *map* of pseudotime *i*nto real-*t*ime). MAPiT makes use of prior knowledge of the distribution of cells on the real scale to derive the requested transformation (Figure 2.1). We demonstrate MAPiT for a temporal scale in cell cycle studies and for a spatial scale representing the distance from the surface in multi-cellular tumour spheroids (MCTS). MAPiT robustly reconstructs the true scale of both processes which we verified with imaging data.

2.2 MAPiT - *MAP* of pseudotime *i*nto real-*T*ime

This section presents the theoretic concept for MAPiT and discusses some practical implications.

2.2.1 Measure-preserving transformation for probability distributions

The theoretic foundation of MAPiT originates from measure and probability theory. MAPiT is based on a "measure-preserving transformation" which ensures that the area under the curve is conserved when transforming a probability distribution. Consider a measure space $(X, \mathcal{L}, \lambda)$, where X is a set, $\mathcal{L}$ is a $\sigma -$ ring of measurable subsets of X, and λ is the

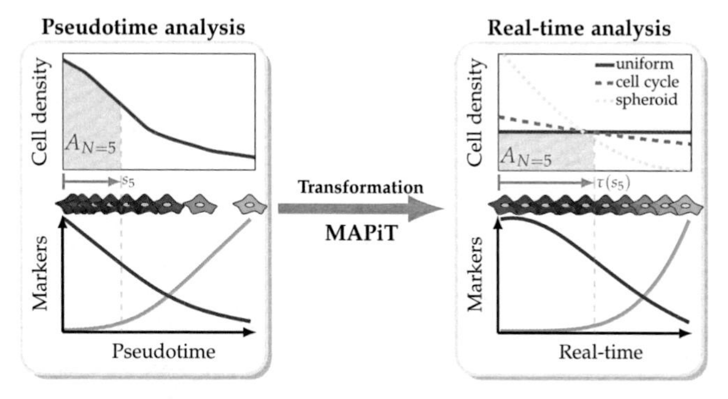

Figure 2.1. MAPiT deduces process dynamics from single-cell snapshot data. Cell density and marker trajectories on pseudotime scale vary with the distance measure used by the pseudotime algorithm and real temporal trajectories cannot be deduced. Cell density, order and trajectories for two markers on pseudotime scale are shown for an exemplary process. As an example pseudotime position of the fifth displayed cell s_5 and associated area under the cell density curve $A_{N=5}$ are indicated in gray. Nonlinear transformation of pseudotime scale recovers true scale dynamics. MAPiT uses prior knowledge of cell densities on the real scale to transform pseudotime to real time by enforcing equality for the area under the density curves at corresponding points on both scales (gray areas). Cell order and marker trajectories are shown for an exemplary uniform distribution on the real scale. Positions of cells across the cell cycle (dashed, orange) or decreasing number of cells towards the center of spheroid cultures (dotted, yellow) are other real scale densities.

measure. Given a map τ from a measure space $(X, \mathcal{L}, \lambda)$ to a measure space $(Y, \mathcal{S}, \mu)$, τ is called measurable if $A \in \mathcal{S}$ implies $\tau^{-1}(A) \in \mathcal{L}$. Given that τ is measurable, τ is called measure-preserving if $A \in \mathcal{S}$ implies $\lambda(\tau^{-1}(A)) = \mu(A)$. We denote the pseudotime values with $s \in [0,1]$, real-time scale with $x \in [0,T]$ and measured signals with $y \in \mathbb{R}$. Based on the general definition of a measure-preserving map τ, the transformation $\tau \colon s \to x$ of a probability density of cells in pseudotime $p_s(s)$ to the probability density of cells on the real-time scale $p_x(x)$ reads

$$p_x(x) = \left| \frac{\mathrm{d}\tau^{-1}(x)}{\mathrm{d}x} \right| p_s\left(\tau^{-1}(x)\right) , \tag{2.1}$$

$$P_x(x) = P_s\left(\tau^{-1}(x)\right) . \tag{2.2}$$

The mapping $\tau : s \to x$ from pseudotime to real-time was obtained by solving Eq. (2.2) for τ, which then depends on the probability mass, or cumulative density, of cells on both scales

$$\tau^{-1}(x) = P_s^{-1}\left(P_x(x)\right) , \text{ or} \tag{2.3}$$

$$\tau(s) = P_x^{-1}\left(P_s(s)\right) . \tag{2.4}$$

Thus, by definition, the transformation τ requires knowledge of the density (or cumulative density) of cells on the pseudotime scale and the desired scale. We know from probability theory, that these distributions are strictly positive over their support. Accordingly, the cumulative distributions are monotonically increasing and the inverses exist. Once the mapping τ is known, one can apply the transformation to the joint probability densities of pseudotime and the observed quantities $p_s(s,y)$ to obtain the desired joint distribution of the true scale x and measured markers y

$$p_x(x,y) = \left| \frac{\mathrm{d}\tau^{-1}(x)}{\mathrm{d}x} \right| p_s\left(\tau^{-1}(x), y\right) . \tag{2.5}$$

Although, requiring knowledge of the true scale distribution may appear to be a strong restriction, this distribution can often be obtained from simple considerations.

2.2.2 Properties of MAPiT

Ways to obtain the target distribution

The transformation requires knowledge of the distributions (or cumulative distributions) of cells on the pseudotime and the desired scale. Pseudotime values from experimental data can be used to calculate the distribution on the pseudotime scale. The distributions on the real scales can either be derived from theoretical considerations or from empirical measurements. We will demonstrate for cell cycle and spheroid reconstruction, how a priori knowledge of the process of interest can be used to derive the distribution of cells on the desired real-time scale.

Error handling

The transformation of pseudotime data to a real scale allows a direct comparison of the inferred trajectories with ground truth data from other experimental methods, such as live cell microscopy. A mismatch between the MAPiT trajectories and ground truth data has two basic reasons (1) wrong pseudotemporal order or (2) wrong distribution of cells on the true scale. A wrong pseudotemporal order can result in qualitative deviations, in the sense that the signs in the slope of the trajectories differ in their sequence. In contrast, a wrong distribution of cells on the true scale can only result in rather quantitative differences such that the slope of the trajectory might be different, but not the sign sequence. Thus, MAPiT can be used to detect erroneous pseudotime ordering.

2.3 Reconstructing spheroids with MAPiT

This section demonstrates the application of MAPiT in order to map pseudotime values to a spatial scale representing the distance from the surface in multi-cellular tumour spheroids. Two-dimensional (2-D) cell-based assays have been the dominant approach used to study cellular processes in human disease and for drug discovery (L. Li and D. LaBarbera 2017). These 2-D cell models are cost-effective and highly amenable to high-throughput and high-content screening. However, their responses and phenotypes only partially represent the pathology of human disease leaving a huge gap between observations in *in vitro* experiments and the true *in vivo* processes (Pampaloni et al. 2007). This gap is one of the major reasons for failure in clinical trails. Organoids and in particular MCTS fill this gap promising improved clinical translation success rates. Therefore, high throughput screenings of MCTS provide great potential for future biomedical research.

However, there are two major barriers in the use of 3-D-spheroid models in a high-throughput and high-content fashion. High-throughput is impaired by the preparation procedure for spheroid samples often requiring experimentally challenging slicing of individual spheroids for imaging. High-content analysis including the spatial position of single-cells on the other hand is limited by the capability of the imaging methods (D. V. LaBarbera et al. 2012). Characterizing the spatial composition of MCTS is based on advanced microscopy like confocal laser scanning microscopy and spinning disc or immunohistochemistry of spheroid cuts (D. V. LaBarbera et al. 2012). Several other approaches exist for single-cell analysis in spatial context either on protein level, like image mass spectrometry (Giesen et al. 2014), or at the RNA level, review in Strell et al. (2019). For example, Arnol et al. (2019) recently exploited spatial variance signatures in spatial gene expression data revealing cell-cell interactions as a major driver of protein expression heterogeneity. Furthermore, computational approaches like Achim et al. (2015) and Satija et al. (2015) employ reference gene expression databases for spatial mapping of single-cell RNA-seq.

Dissociation of MCTS in single-cells causes loss of spatial information, but allows high throughput multiplex measurements like flow cytometry, CyTOF or scRNA-seq. Figure 2.2 illustrates the workflow for recovering spatial information in single-cell data from dissociated spheroids.

2.3.1 Problem formulation

Multicellular spheroids grown from cancer cells are widely used as avascular tumour models (Vörsmann et al. 2013; Jabs et al. 2017; Lin and Chang 2008). As a consequence of nutrient and oxygen deprivation within the spheroids, proliferative cells begin to enclose inner layers

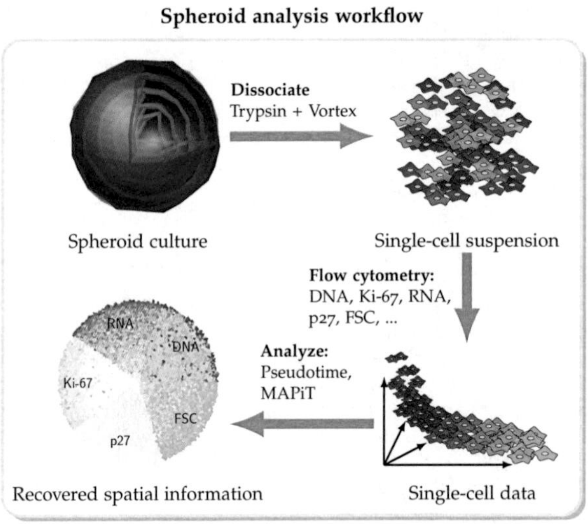

Figure 2.2. MAPiT recovers spatial positions of cells within spheroids from flow cytometric data. Illustration of spheroid analysis workflow. Individual cells derived from dissociated spheroids are analysed for different markers by flow cytometry. Spatial information can be recovered by applying MAPiT to pseudotime trajectories of measured markers.

of quiescent and necrotic cells, resembling a zonation found in solid tumours (Freyer and Sutherland 1986; D. V. LaBarbera et al. 2012). Current routine methods to study spatial distributions and patterns of cellular markers are restricted to intact spheroids. Experimental analysis of intact spheroids is technically cumbersome and of limited throughput, since available methods rely on sequential spheroid fixation, sectioning and imaging procedures. By dissociating tumour spheroids for single-cell experiments, spatial information across which cell-to-cell heterogeneities in tumour cell spheroids manifest is lost. We applied MAPiT to study if we can recover spatial scales from flow cytometric measurements of dissociated spheroids in a reliable and robust manner.

For our studies, we grew spheroids of HCT116 cells to diameters of approximately 500 µm. Besides standard readouts like forward scatter (FSC), indicating cell volume, cells were in addition stained for DNA as measure for cell cycle stage, RNA as indicator for transcriptional activity, Ki-67 as marker for proliferation and p27 as marker for quiescence (Figure 2.2). The pseudotime obtained by applying pseudotime algorithms represents the sequence of changes in the abundance of cellular components from the rim to the core, from the proliferative to the necrotic zone, respectively. Thus, we define the problem for spheroid analysis as follows:

Problem 2.2. Given single-cell data and pseudotime values from dissociated spheroids, recover their spatial position of individual cells.

2.3.2 Distribution on desired scale

To apply MAPiT to these data, the distribution of cells on the spatial scale had to be taken into account, which in the case of radial symmetry of spheroids is the density of cells in relation to the distance from the spheroid surface. Cell density depending on the distance from the surface was obtained from sphere geometry (Figure 2.3). Our analysis is based on

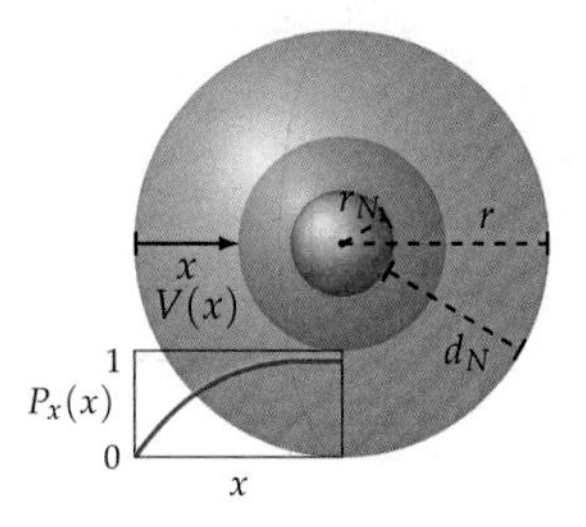

Figure 2.3. Sphere geometry. Volume of spherical shell $V(x)$ (light blue) with radius r and thickness x, normalized to the volume of the sphere $V_S(r)$ (blue) minus the volume of the necrotic core $V_S(r - r_N)$ (gray) provides the cumulative distribution $P_x(x)$ of cells in the spheroid.

the following assumptions: (1) Spheroids are radial symmetric, (2) all cells in the spheroid have the same size. We verified both assumptions by visual inspection of whole spheroids and spheroid slices. We can thereby derive the density of cells in a spheroid as follows: The volume of a sphere with radius r is given by

$$V_S(r) = \frac{4}{3}\pi r^3 . \tag{2.6}$$

The volume of a spheroid with necrotic core with radius $r_N = \max(r - d_N, 0)$ equals

$$V_M(r, r_N) = V_S(r) - V_S(r_N)\,, \tag{2.7}$$

with d_N being the distance from the surface where the necrotic core begins. The volume of a spherical shell at distance x from the surface of the spheroid with necrotic core is then given by

$$V(x) = V_M(r, r_N) - V_M(r - x, r_N)\,. \tag{2.8}$$

Normalising Eq. (2.8) with the total spheroid volume V_M results in the normalised volume with respect to the distance to the surface of the spheroid which represents the cumulative distribution of cells related to the distance from the surface

$$P_x(x) = \frac{V(x)}{V_M(r, r_N)} = \frac{r^3 - (r-x)^3}{r^3 - r_N^3}\,. \tag{2.9}$$

MAPiT is furthermore based on the probability density function and the inverse of the cumulative distribution which can be calculated analytically

$$p_x(x) = 3\frac{(r-x)^2}{r^3 - r_N^3}\,, \tag{2.10}$$

$$P_X^{-1}(y) = r - \sqrt[3]{y\,(r_N^3 - r^3) + r^3}\,. \tag{2.11}$$

Spheroid radius r and the radius of first appearance of a necrotic core r_N were inferred from experiments, described in more detail in Appendix C. For the present study we used $r_N = 270$ µm and a time- (t in [days]) dependent spheroid radius based on the linear regression

$$r(t) = 19.2 + 22.4\,t \tag{2.12}$$

for spheroid growth.

Empirically, one could obtain the distance-dependent distribution of cells from images of spheroid slices (for example Figure 2.4 a). Kernel density estimation of the distances of the cells from the surface results in the distribution, or cumulative distribution, of cells in a spheroid slice, in the case of radial symmetry. The mapping $G(x)$ from a 2-D cumulative distribution to the cumulative distribution in a 3-D sphere is calculated based on the geometry of circles (normalised area) and spheres (normalised volume) by

$$P_{3D}(x) = G(x)\,P_{2D}(x)\,, \tag{2.13}$$

with

$$G(x) = \frac{P_{3D}(x)}{P_{2D}(x)} = \frac{r^3 - (r-x)^3}{r^3 - r_N^3}\left(\frac{r^2 - (r-x)^2}{r^2 - r_N^2}\right)^{-1}\,. \tag{2.14}$$

Applying this mapping to an empirical 2-D distribution of cells results in the 3-D distribution of cells which can be used in MAPiT.

2.3.3 Studying spheroid composition with MAPiT

Here, we present the performance and limitations of studying spheroid composition with MAPiT.

Validation against microscopy images

First, we validate MAPiT against microscopy data. To this end, we obtained pseudotemporal ordering of dissociated cells with wanderlust. Cells with high Ki-67 and high RNA signals are located at the rim of the spheroid and were set as root cells for the algorithm. MAPiT indeed recovered the spatial positions of single cells, with the reconstructed Ki-67 distribution correlating excellently with intensity profiles obtained from confocal microscopy of spheroid sections stained for Ki-67 (Figure 2.4).

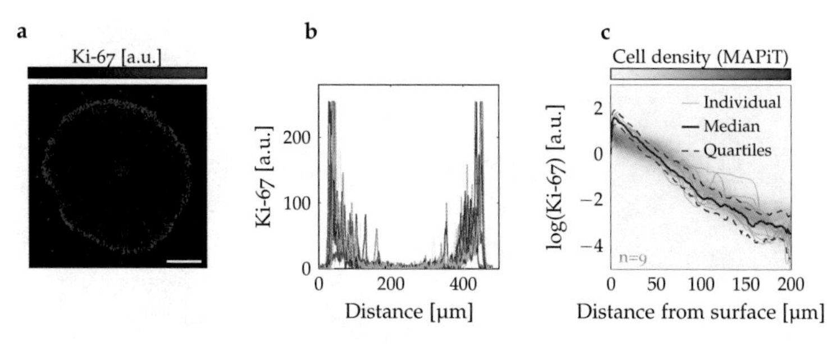

Figure 2.4. MAPiT derived Ki-67 distribution in an 11-day-old HCT116 cell spheroid. (**a**) Representative spheroid cross-section stained for Ki-67. Scale bar: 100 µm. (**b**) Transversal quantified Ki-67 intensities in spheroid sections (rectangles in (a), $n = 9$). (**c**) Ki-67 intensities obtained by MAPiT from flow cytometric data (blue) closely match Ki-67 intensity profiles determined microscopically in spheroid cross-sections.

Capturing the heterogeneity

Next, we examined if MAPiT can also capture the heterogeneity within the different zones in the spheroid, typically a proliferative layer followed by quiescent and finally necrotic cells. The DNA content of cells in the outermost spheroid layers exhibited a bimodality typically for proliferating cells with subpopulations in G1 (2N), S and G2/M (4N) phases (Figure 2.5). In contrast, the majority of cells in the inner layers remain in a quiescent G1/G0 (2N) state. The distinct distributions of additional markers, Ki-67, p27 and RNA content corresponded to this pattern (Figure 2.5).

MAPiT is robust

We then studied the robustness of MAPiT by evaluating its performance to recover the spatial position using different pseudotime algorithms, markers and spheroid sizes. We first used the Wanderlust and Diffusion Maps algorithms (Bendall et al. 2014; Haghverdi et al. 2014) to obtain a pseudotemporal ordering. The order of cells was largely conserved in both algorithms (Figure 2.6, monotonously increasing data points), however the pseudotime values were clearly different between both algorithms (Figure 2.6, deviation from the diagonal).

Transforming the diffusion pseudotime (DPT) and Wanderlust axis to a distance scale with MAPiT resulted in almost identical spatial profiles of Ki-67, RNA and p27 (Figure 2.7),

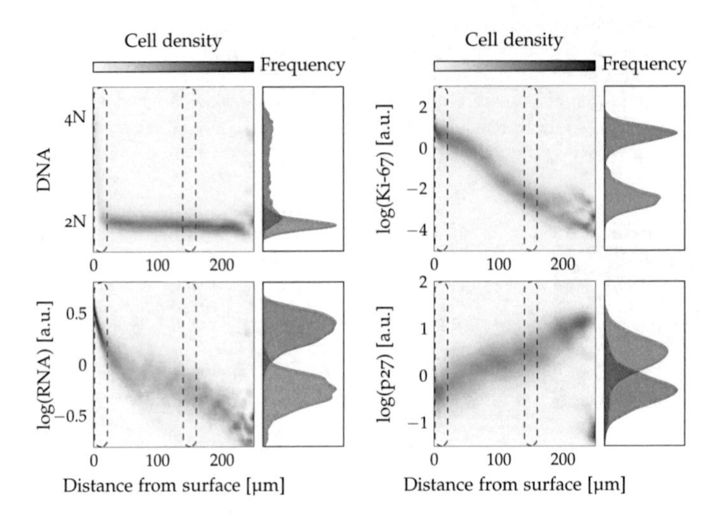

Figure 2.5. Conditional signal densities $p(y|x)$ of indicated markers related to the distance from the surface, as obtained by MAPiT. Signal frequencies at the outermost layer and at 150 μm distance from the spheroid surface, as indicated by the dashed rectangles, are shown for an exemplary 11-day-old spheroid.

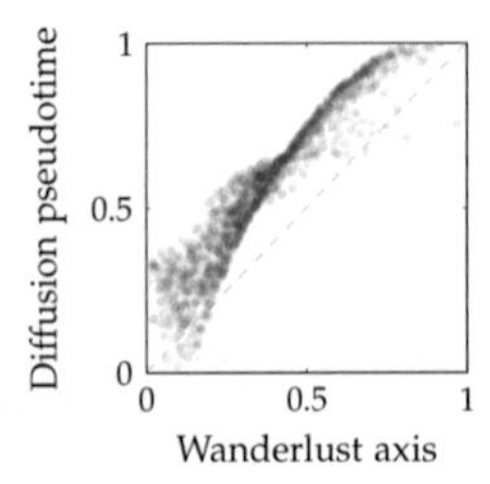

Figure 2.6. Pseudotime values vary for different algorithms.

thereby demonstrating the independence of MAPiT to the pseudotime algorithm. MAPiT should likewise provide identical marker trajectories independent of the choice of markers used to generate the pseudotime order. To validate this, we conducted "leave one out cross validation", where we took subsets of the markers as input to the Wanderlust algorithm, and compared the results obtained by MAPiT. MAPiT provided correct distance profiles of all markers in all combinations, demonstrating its independence of the choice of inputs to the pseudotime algorithms (Figure 2.8). Thus, MAPiT is independent of the choice of markers or pseudotime algorithms, provided that the order of cells on the pseudotime scale reflects the sequence or directionality of the biological processes.

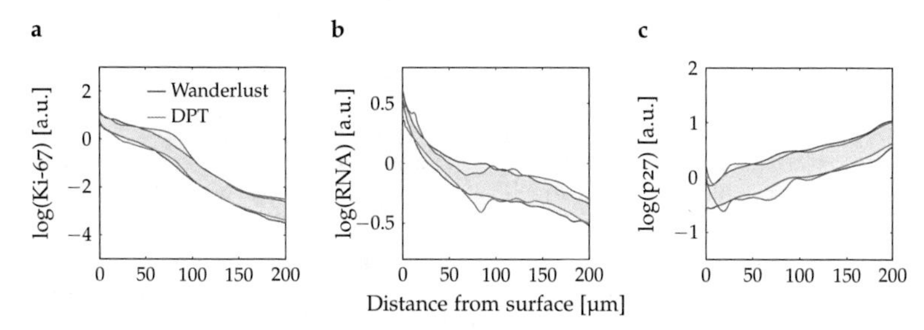

Figure 2.7. MAPiT robustly reconstructs cell positions, irrespective of the pseudotime algorithm used. Ranges show 50% confidence intervals of the (**a**) Ki-67, (**b**) RNA and (**c**) p27 signal intensities obtained from transforming Wanderlust and DPT pseudotime data.

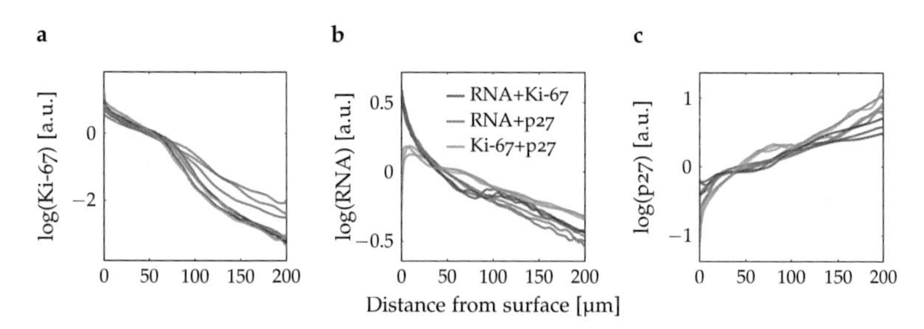

Figure 2.8. Robust reconstruction of distance-dependent signal intensities from observation subsets in 11-day-old HCT116 spheroids. Three replicates of (**a**) Ki-67, (**b**) RNA and (**c**) p27 median intensities of the transformed signal obtained by performing Wanderlust.

Spheroid composition is conserved over different sizes

Spatial reconstruction by MAPiT also provides scope to significantly accelerate high-throughput studies that make use of advanced 3-D culture-based cell screens, such as spheroid based viability assays and drug effect screens. Indeed, spheroids are considered superior models in comparison to conventional cell cultures (D. V. LaBarbera et al. 2012), but spatiotemporal analyses still require cumbersome and manual labour-intensive work for the analysis of cross-sectional slices. We next demonstrate the capability of MAPiT towards high-throughput analysis by examining the composition of spheroids at different diameters. For this purpose, we applied MAPiT to three independent replicates of flow cytometry data obtained from dissociated spheroids at three different diameters of approximately 350 µm, 530 µm and 665 µm corresponding to culture duration of 7, 11 and 14 days, respectively (Figure 2.9 a). MAPiT performed reliably in 3-D reconstruction when processing flow cytometry data from dissociated spheroids of different sizes and ages, based on Ki-67, RNA and p27 amounts (Figure 2.9). The profiles for Ki-67, RNA and p27 were conserved throughout spheroid sizes, indicating a dependence of these markers solely on the distance from the surface (Figure 2.9 c) and therefore their suitability for spatial reconstruction.

Changes observed for p27 at day 7 however were less pronounced than the changes observed at days 11 and 14. This difference might be due to the emergence of a necrotic core in spheroids with diameters larger 500 µm present at culture duration of 11 and 14 days, but not 7 days. The emergence of necrotic cells might cause additional gradients, thereby influencing the heterogeneity present in MCTSs and tumor tissue. The pronounced change of the markers close to the spheroid core is caused by outliers which are often located at the very end of the pseudotime scale.

The data in this experiment was obtained by an automatable procedure comprising spheroid growth, harvesting, dissociation, staining, data collection and subsequent analysis with MAPiT. Current routine methods would in contrast require non-automatable and elaborate sectioning of the spheroids, thereby preventing high-throughput studies.

In summary, robust spatial markers together with MAPiT therefore allow the rapid and versatile reconstruction of spheroids, providing a basis for studying spatiotemporal changes in other measurable variables.

2.4 Analysing cell cycle progression with MAPiT

This section demonstrates the application of MAPiT for a temporal scale representing the age of cells in their cell cycle. Continuous cyclic processes such as the cell cycle are of particular importance in cell biology. The cell cycle itself is one of the fundamentals of life in general and knowledge about its regulation is crucial in the treatment of various diseases, most prominently cancer. The molecular machinery which drives and regulates the cell cycle progression has been rigorously investigated and is well understood (Kapuy et al. 2009; Morgan 2007; Norbury and Nurse 1992). Cell cycle-dependent variability in cellular activities, such as stimulus response or in general cellular signalling, though, is poorly understood and remains experimentally challenging.

There are three methods which are mainly used to study the cell cycle: (1) single-cell methods where individual cells are analysed over time, (2) analysis of a synchronized cell population over time and (3) analysis of snapshot population data with mathematical methods to extract dynamic information of the process. Tracking individual cells using time-lapse microscopy is in many situations not possible or impractical. Obviously, destructive methods such as immunofluorescence, flow cytometry or single-cell sequencing are not feasible. For example, fluorescent live-cell reporters are necessary for the analysis of protein abundances. The synchronization of an asynchronous cell population by environmental or chemical treatments may be accompanied by complex side effects which may affect measurements

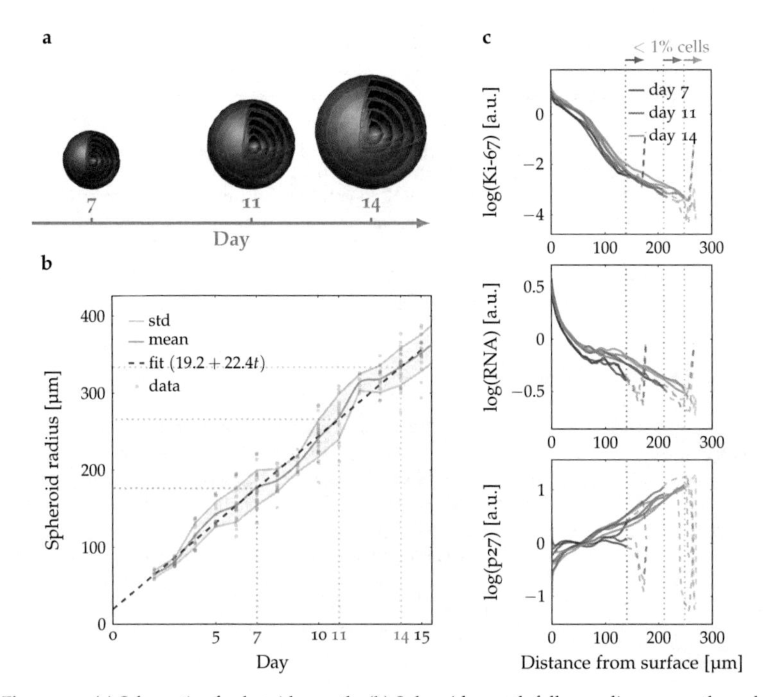

Figure 2.9. (**a**) Schematic of spheroid growth. (**b**) Spheroid growth follows a linear growth model. Microscopically obtained data shows spheroid radius over 15 days after seeding from $n = 23$ spheroids in three independent replicates. (**c**) Median profiles for Ki-67, RNA and p27, as obtained by MAPiT, were conserved throughout spheroid sizes ($n = 3$). Notable deviations towards the ends of the profiles (dashed) and thus within the center of the spheroids arise from inaccuracies due to the low number of cells available for analysis at these locations.

obtained with standard molecular biology experimental protocols such as western blots or flow cytometry (Banfalvi 2011; Sigal et al. 2006; Xeros 1962) and bulk transcriptional analysis with RNA-seq (Y. Liu et al. 2017). Furthermore, the temporal accuracy of this method is limited as a synchronized population often rapidly looses synchrony.

The third, less common method exploits ergodic principles to infer the rates of cell cycle-dependent molecular events, such as biochemical reactions from population snapshot experiments (Kafri et al. 2013; P. Liu et al. 2014; Wheeler 2015; Kuritz et al. 2017; Kuritz et al. 2020b). The approach, based on ergodic theory, uses the fact that the proportion of cells in different stages of the cell cycle does not change in an asynchronous population. In such a steady state population the number of cells in a particular stage is related to the average transit time through that stage. Additional knowledge of the population growth rate makes it possible to infer cell cycle-dependent rates on the level of the cell population at any point of the cell cycle. We present this approach and further extensions to noisy progression in Chapter 3. In the following we present the application of MAPiT to recover temporal dynamics across the cell cycle from static single-cell measurements. After formulating the problem we derive the real-scale distribution and elucidate the assumptions which we pose on the experimental setup. Finally, we apply MAPiT to snapshot data from a flow-cytometric measurement of cell cycle markers in NCI-H460 cells and compare the results with single-cell trajectories from a time-lapse imaging experiment.

2.4.1 Problem formulation

For the analysis of cell cycle progression in single-cell snap-shot data, we used a static flow cytometric measurement of DNA and *mAG-hGeminin (1-110)*, a fluorescent ubiquitination-based cell cycle indicator (Figure 2.10 a) (Sakaue-Sawano et al. 2008; Kuritz et al. 2017; Kuritz et al. 2020b), in unperturbed NCI-H460/geminin cells to reconstruct the kinetics of geminin. The pseudotime obtained with the markers (Figure 2.10 b) should be mapped to the temporal scale on which the cell cycle progresses, namely the age of a cell (time since cytokinesis):

Problem 2.3. Given single-cell data and pseudotime values from an unsynchronised cell population in its exponential growth phase, recover the temporal dynamics of single cells in their cell cycle.

2.4.2 Distribution on desired scale

For analysis of cell cycle-dependent processes with single-cell measurements, MAPiT requires the distribution of cells related to their cell cycle stage or equivalently their age a. Cell age refers to the time since cell birth via cytokinesis. Our analysis is based on the following assumptions: (1) population is unperturbed and in its exponential growth phase, (2) no cell death in the population, (3) cell cycle progression is homogeneous. We thus restricted our analysis to unperturbed cell populations in their exponential growth phase with growth rate γ and cell cycle length T related by $\gamma = \frac{\ln 2}{T}$. In such a case, the theoretical steady state age distribution of a cell population is given by (Powell 1956) (Figure 2.10 c):

$$p_x(x) = \frac{2 \ln 2}{T} 2^{-x/T} \,. \tag{2.15}$$

The cumulative distribution and its inverse can be obtained in closed form:

$$P_x(x) = 2\left(1 - 2^{-x/T}\right), \tag{2.16}$$

$$P_x^{-1}(y) = -T\log_2\left(1 - \frac{y}{2}\right). \tag{2.17}$$

Thus, in case of an unperturbed cell population it is sufficient to know the growth rate of the population to recover cellular age with MAPiT and thereby obtain the temporal changes related to cell cycle progression of measured markers from one single-cell experiment. We verified all assumptions by live cell imaging. In addition, light scattering characteristics in the flow cytometric data sets were used to probe the population for cell death. Empirically, the distribution of cells in their cell cycle can for example be obtained by live cell microscopy. The age distribution in the population after recording for at least one cell cycle represents the desired distribution. Density in pseudotime for cell cycle data was estimated by kernel density estimation with linked boundary conditions to account for doubling of cell density during cell division (Colbrook et al. 2020).

2.4.3 Cell cycle trajectories obtained by MAPiT match live-cell microscopy

Temporal trajectories of geminin obtained with MAPiT corresponded excellently to geminin kinetics obtained by single-cell time-lapse microscopy (Figure 2.10 d), as exemplified by the boost in geminin intensity at approximately 7 h, indicating the onset of S-phase. This result highlights the temporal accuracy of the real-time scale obtained by MAPiT for cell cycle analysis based on snapshot flow cytometric data.

2.5 Summary and discussion

Summary

In summary, MAPiT provides a solution to a fundamental problem, namely the transformation of pseudotime to real-time or the true scale of a biological process. Based on ergodic analysis, MAPiT uses the fact that the number of cells in a particular stage is related to the average transit time through that stage, allowing inference of complex dynamics from a single population measurement.

For example, snapshot single-cell data of cell populations can be converted to extract real-time kinetics of cellular processes and responses, which otherwise could only be obtained by live-cell microscopy, which is more complex, time consuming and limited by the availability of suitable live-cell reporters. By reconstructing spatial and temporal spheroid compositions from single-cell data, MAPiT provides insights to the evolution of cellular heterogeneity within tumour-like microenvironments and allows to understand how responsiveness to therapeutics manifests within spheroidal environments.

Comparison with other methods

To allow for a profound discussion of MAPiT and other methods, we briefly review the underlying mathematical description of 1D processes related to the transformation of pseudo-

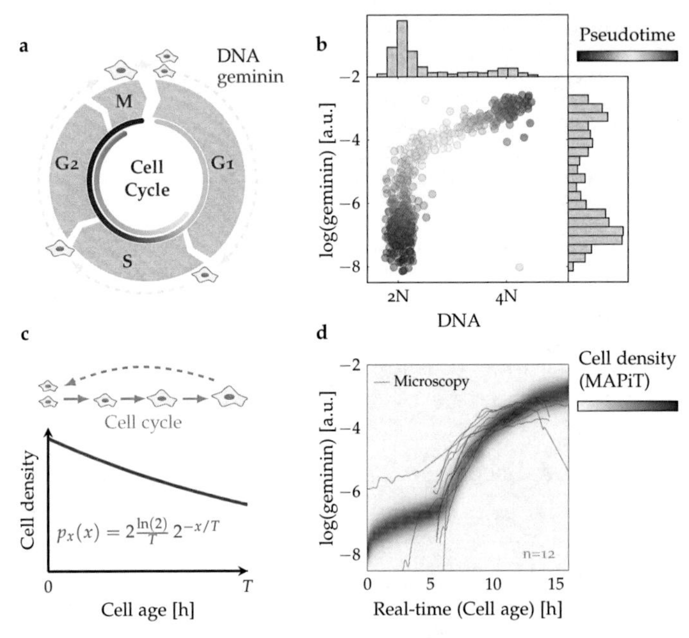

Figure 2.10. MAPiT recovers cell cycle dynamics. (**a**) Schematic of the cell cycle with geminin expression, a marker for cell cycle progression, starting at the onset of S phase. (**b**) DNA and geminin signals from an unsynchronised population of NCI-H460/geminin cells were used to obtain a pseudotemporal ordering of the population. (**c**) MAPiT employs steady state age distribution of an unsynchronised cell population. (**d**) Reconstructed temporal profile of geminin signal density obtained with MAPiT and single-cell trajectories from time-lapse imaging correlate strongly.

time to real-time (see also Chapter 3). The evolution of the number density $n(s,t)$ of cells on the process manifold $s \in [0,1]$ can be described by a second order partial differential equation

$$\frac{\partial}{\partial t}n(s,t) + \frac{\partial}{\partial s}\left(v(s)n(s,t)\right) - \frac{1}{2}\frac{\partial^2}{\partial s^2}\left(g(s)^2 n(s,t)\right) = \left(\alpha(s) - \beta(s)\right) n(s,t)\,, \tag{2.18}$$

with position-dependent velocity $v(s)$ and diffusion $g(s)$. Sources $\alpha(s)$ and sinks $\beta(s)$ capture growth of the population, for example, by cell division and vanishing cells, for example, due to cell death. Neumann boundary conditions describe influx and outflow at the beginning and end of the process

$$\begin{aligned} J_0(t) &= \frac{\partial}{\partial s}n(0,t) \\ J_1(t) &= \frac{\partial}{\partial s}n(1,t)\,, \end{aligned} \tag{2.19}$$

and an initial condition for the distribution at time $t = 0$ completes the model

$$n(s,0) = n_0(s)\,. \tag{2.20}$$

Few methods allow the transition from pseudotime to real-time analysis: *Pseudodynamics* infers real-time dynamics by estimating the distribution of a cell population across a continuous cell state coordinate over time based on a stochastic differential equation (Fischer et al. 2019). By using a regularisation Fischer and colleagues try to overcome the non-identifiability of velocity and combined entry and exit rate. However, estimated functions may not be uniquely identifiable and the method is computationally demanding which might impair its applicability.

The method called *velocyto* uses a different approach (La Manno et al. 2018), assuming that the balance between unspliced and spliced mRNAs is predictive of cellular state progression. Thereby, local velocity is inferred from the abundance of splicing variants at each point in the scRNA-Seq data space. The concept is very elegant, however the method is restricted to single-cell transcriptomics and the assumption of constant rates might not always hold. Bergen et al. (2019) recently proposed a generalised framework (*scVelo*) based on the same concept as *velocyto*.

Moreover, as we describe in Chapter 3, ergodic theory on cyclic processes is used to infer the rates of molecular events during cell cycle progression from single-cell measurements (Kafri et al. 2013; Kuritz et al. 2017).

Haghverdi and colleagues suggested inference of a universal time by integrating over the inverse distribution in pseudotime (Haghverdi et al. 2016). The underlying theory is based on the fact that the density on the pseudotime scale is inversely correlated with the velocity, with which the cells progress through this position on the process manifold. Given the partial differential equation Eqs. (2.18) to (2.20), this holds true if and only if

$$\alpha(s) - \beta(s) = 0\,, \tag{2.21}$$

$$g(s) = 0\,, \tag{2.22}$$

$$J_0(t) = J_1(t)\,. \tag{2.23}$$

This is equal to assuming that, (1) population growth (by cell division) and cell removal (by cell death) balance each other, (2) all cells are identical and (3) influx of cells in the beginning of the process is equal to the outflow at the end. While (2) and (3) are under certain conditions reasonable assumptions, balancing of population growth and cell death (3) at each stage of the process is only justified in special cases. However, if (1)-(3) hold and the process is in steady state operation, then one can deduce, based on ergodic principles, that the derived universal time is the same as the real-time.

In another approach by Weinreb and colleagues, population balance analysis was used to infer diffusion $g(s)$ as well as entry and exit rates $\alpha(s), \beta(s)$ in a PDE similar to Eqs. (2.18) to (2.20) (Weinreb et al. 2018). However, the recovered time scale was still not identical to real-time due to non-identifiability of velocity and combined entry and exit rates at the same time.

Discussion and Outlook

MAPiT decouples pseudotime algorithms and observations from the actual time of the process which facilitates direct fusion of data from single-cell snapshot experiments with other methods such as live cell microscopy, thereby bringing experimental results to a unified scale. By extension, applying MAPiT to other single-cell snapshot data, such as single-cell transcriptomics and proteomics data, might significantly improve the inference of complex regulatory processes and networks by recovering real temporal and spatial dynamics. Furthermore, MAPiT may allow direct inference of gene regulatory networks (GRNs) from snapshot data which was so far a difficult task due to the missing real-time axis (Papili Gao et al. 2017). Recently, pseudotime algorithms were further developed to robustly recognise also branching processes in differentiation pathways (Haghverdi et al. 2016; Setty et al. 2016; Qiu et al. 2017), providing scope to apply MAPiT to study differentiation dynamics in individual branches.

Overall, MAPiT is a robust and universal tool to recover temporal or spatial cellular trajectories from high-throughput, high-dimensional single-cell experiments. MAPiT can be combined with pseudotime algorithms, and a MATLAB implementation is available through GitHub (Kuritz 2020, `https://doi.org/10.5281/zenodo.3630379`).

Chapter 3

Cell cycle analysis with ergodic principles and age-structured population models

This chapter is based on the publication:

Karsten Kuritz et al. (2017). 'On the relationship between cell cycle analysis with ergodic principles and age-structured cell population models'. In: *Journal of Theoretical Biology* 414, pp. 91–102. DOI: 10.1016/j.jtbi.2016.11.024.

Cyclic processes, in particular the cell cycle, are of great importance in cell biology. Continued improvement in cell population analysis methods like fluorescence microscopy, flow cytometry, CyTOF or single-cell omics made mathematical methods based on ergodic principles a powerful tool in studying these processes. In this chapter, we establish the relationship between cell cycle analysis with ergodic principles and age-structured population models. To this end, we describe the progression of a single cell through the cell cycle by a stochastic differential equation on a one-dimensional manifold in the high-dimensional dataspace of cell cycle markers. Given the assumption that the population is in a steady state, we derive transformation rules which transform the number density on the manifold to the steady state number density of age-structured population models. Our theory facilitates the study of cell cycle-dependent processes including local molecular events, cell death and cell division from high-dimensional snapshot data. Ergodic analysis can in general be applied to every process that exhibits a steady state distribution. By combining ergodic analysis with age-structured population models this chapter provides the theoretic basis for extensions of ergodic principles to distribution that deviate from their steady state.

The experimental data that we present in this chapter was prepared by the Morrison Lab at the Institute of Cell Biology and Immunology at the University of Stuttgart. This chapter is taken in parts from Kuritz et al. (2017).

3.1 Background and problem formulation

Continuous cyclic processes such as the cell cycle are of particular importance in cell biology. The cell cycle itself is one of the fundamentals of life in general and knowledge about its regulation is crucial in the treatment of various diseases, most prominently cancer. The molecular machinery which drives and regulates the cell cycle progression has been rigorously investigated and is well understood (Kapuy et al. 2009; Morgan 2007; Norbury and Nurse

1992). Cell cycle-dependent variability in cellular activities, such as stimulus response or in general cellular signalling, though, is poorly understood and remains experimentally challenging.

Ergodic principles have been used to infer the rates of cell cycle-dependent molecular events, such as biochemical reactions from population snapshot experiments (Kafri et al. 2013; P. Liu et al. 2014; Wheeler 2015). The approach termed ergodic analysis (EA) uses the fact that the proportion of cells in different stages of the cell cycle does not change in an asynchronous population. In such a steady state population the number of cells in a particular stage is related to the average transit time through that stage. Additional knowledge of the population growth rate makes it possible to infer cell cycle-dependent rates on the level of the population at any point of the cell cycle.

In this chapter we introduce the relation between EA-derived molecular rates and age-structured population models. Our theory establishes the connection between cell cycle-dependent signalling captured by EA and cellular events such as cell death and cell division which are represented in age-structured population models. The approach is based on the description of the progression of a single cell through the cell cycle in an arbitrary parameterisation by a stochastic differential equation (SDE). This SDE retrieves in its simplified form where the noise is omitted previous results described in Section 2.4 and Kafri et al. (2013) and provides accurate cell cycle-stage specific rates for molecular events. By accounting for the noise during the cell cycle progression we additionally capture the variability present within a population. We thus arrive at the following problem formulation:

Problem 3.1. Given single-cell data and pseudotime values from an unsynchronised population in its exponential growth phase and the description of the process as stochastic differential equation we aim at identifying the relation of the position of a cell in data space to its age.

This chapter is structured as follows: First we introduce the class of age-structured populations models. In Section 3.3 we establish the relationship between ergodic analysis and age-structured population models: we define the SDE model for single cell progression through the cell cycle which will be transformed to a steady state age-structure, first without considering noise and second with noisy progression. Different noise sources are covered by differentiating between two different types of noise: (1) cell cycle progression with additive noise from a Wiener process and (2) noise present as distributed cell cycle progression rate in the population. The derived transformations are then applied to a flow cytometry dataset of cell cycle markers and compared with results from time-lapse live-cell microscopy experiments in Section 3.4.

3.2 Age-structured population models

Structured population models have a long history which originated from questions in ecology (McKendrick 1926; Sharpe and Lotka 1911; Foerster 1959). The model class of age-structured population models is based on the well known von Foerster-McKendrick models and are widely used to study cell cycle-related processes (Billy et al. 2013; J. Clairambault et al. 2009; Jean Clairambault et al. 2011; Gabriel et al. 2012; Gyllenberg and Webb 1990; Hross and Hasenauer 2016). We here define the cellular age $a \in \mathbb{R}_{\geq 0}$ that we will refer to in the remainder of the chapter as the time that a cell was alive since its origin from the last cell division. Thus, two daughter cells with age zero emerge from the division of a mother cell. Age-structured population models describe the time $t \in \mathbb{R}$ evolution of the age-dependent cell number density $n(a,t)$ as first order partial differential equation (PDE)

$$\frac{\partial n(a,t)}{\partial t} + \frac{\partial n(a,t)}{\partial a} = -\left(\alpha(a,t) + \beta(a,t)\right) n(a,t). \tag{3.1}$$

The evolution of the number density is dependent on the time- and age-dependent division rate $\alpha(a,t)$ and death rate $\beta(a,t)$. Boundary and initial conditions for newborn cells and the initial age structure respectively of the form

$$n(0,t) = 2\int_0^\infty \alpha(\tilde{a},t) n(\tilde{a},t)\,\mathrm{d}\,\tilde{a}, \tag{3.2}$$

$$n(a,0) = n_0(a) \tag{3.3}$$

complete the model. We will use the following properties of the model given by Eqs. (3.1) to (3.3) in the main part of the article:

The number of cells in the population at any time point is given by the integral over all ages of the population density

$$N(t) = \int_0^\infty n(\tilde{a},t)\,\mathrm{d}\,\tilde{a}. \tag{3.4}$$

The age distribution $p_a(a,t) = \frac{n(a,t)}{N(t)}$ of an exponentially growing cell population with a death rate that is age independent $\beta(a,t) = \beta(t)$ and a growth rate that is time independent $\alpha(a,t) = \alpha(a)$ converges to a steady state distribution $p_a(a)$ which was shown using semi-group methods (Metz and Diekmann 1986) or general relative entropy techniques (Gabriel et al. 2012; Perthame and Zubelli 2007)

$$p_a(a) = 2\gamma e^{-\gamma a} e^{-\int_0^a \alpha(\tilde{a})\,\mathrm{d}\,\tilde{a}}. \tag{3.5}$$

The constant γ corresponds to the growth rate of the population. The age distribution in the steady state simplifies to

$$p_a(a) = \begin{cases} 2\gamma e^{-\gamma a}, & a \leq \ln(2)/\gamma \\ 0, & a > \ln(2)/\gamma \end{cases} \tag{3.6}$$

if all cells in the population undergo division at the same age $T = \ln(2)/\gamma$ (Powell 1956). The division rate is then given by a Dirac delta function at the division age $\alpha(a) = \delta(a - T)$. In the next sections, we provide transformation rules which bring the cell number densities obtained from the EA method to the steady state age-structure of a population.

3.3 Relationship between ergodic analysis and age-structure population models

The EA method exploits the fact that the number of cells in a particular state is related to the average transit time through that state. In the following we propose a model which describes the progression of a single cell through the cell cycle in an arbitrary parameterisation by a stochastic differential equation. This model provides in its simplified form where the noise term is zero an identical progression rate for cells on the cell cycle as derived in Kafri et al. 2013. Furthermore we prove that in this case the transformation to an age parameterisation of the cell cycle results in the steady state age structure given in Eq. (3.6). This observation validates the results in Chapter 2 where we presented the mapping between pseudotime scale and real scales based on knowledge of the distribution on the real scale.

After presenting the transformation without noise, we deduce a transformation which also accounts for noise during progression through the cell cycle and therefore results in the age-dependent number density in its general form as stated in Eq. (3.5). As mentioned in the introduction, one often differentiates between intrinsic and extrinsic noise. Intrinsic noise is characterized by the absence, or only short time correlation between quantities in identical cells of a population. Extrinsic noise on the other hand shows long time correlations of the quantities in a population (Elowitz et al. 2002; Swain et al. 2002; Munsky et al. 2009; Iversen et al. 2014). Such persistent variance between cells in a population is caused by various factors including the local environment or the history of cells (Snijder et al. 2009; Gut et al. 2015; Sandler et al. 2015). The description of noisy cell cycle progression by a stochastic differential equation in Section 3.3.3 does, due to the nature of the Wiener process, not exhibit any correlation. We thus refer to this type of noisy cell cycle progression as intrinsic noise. In Section 3.3.4 we model the noise in cell cycle progression in a population with a distributed progression rate without any stochastic contribution. The variance in the population is then fully correlated throughout the cell cycle and we thus refer to this type of noise in cell cycle progression as extrinsic noise. We provide an overview of the methods in Figure 3.1.

3.3.1 Cell cycle progression model and derivation of the velocity equation

We define the progression of a single cell through the cell cycle on a one dimensional manifold parameterized with $s \in [0,1]$ by an Itô-type stochastic differential equation. This may for example be any pseudotime scale, obtained by trajectory inference method as describe in Section B.1. Furthermore, a reflective boundary at $s = 0$ together with an absorbing boundary at $s = 1$ prevent back-flux of cells over the division point (Skorokhod 1961; Feller 1954). The velocity of a cell on s is given by the function $f(s)$ and the noise from a Wiener-Process enters with the function $g(s)$

$$\mathrm{d}s = f(s)\,\mathrm{d}t + g(s)\,\mathrm{d}W. \tag{3.7}$$

A partial differential equation can then be used for the population level discription. The Fokker-Planck-Equation describes the time evolution of the number density of cells,

$$\frac{\partial}{\partial t}n(s,t) - \frac{\partial}{\partial s}J(s,t) = 0, \tag{3.8}$$

with the flux $J(s,t)$ based on the SDE as

$$J(s,t) = -f(s)\;n(s,t) + \frac{1}{2}\frac{\partial}{\partial s}\left(g(s)^2 n(s,t)\right). \tag{3.9}$$

The reflecting and absorbing boundaries are represented as

$$J(0,t) = 0, \tag{3.10}$$

$$n(1,t) = 0, \tag{3.11}$$

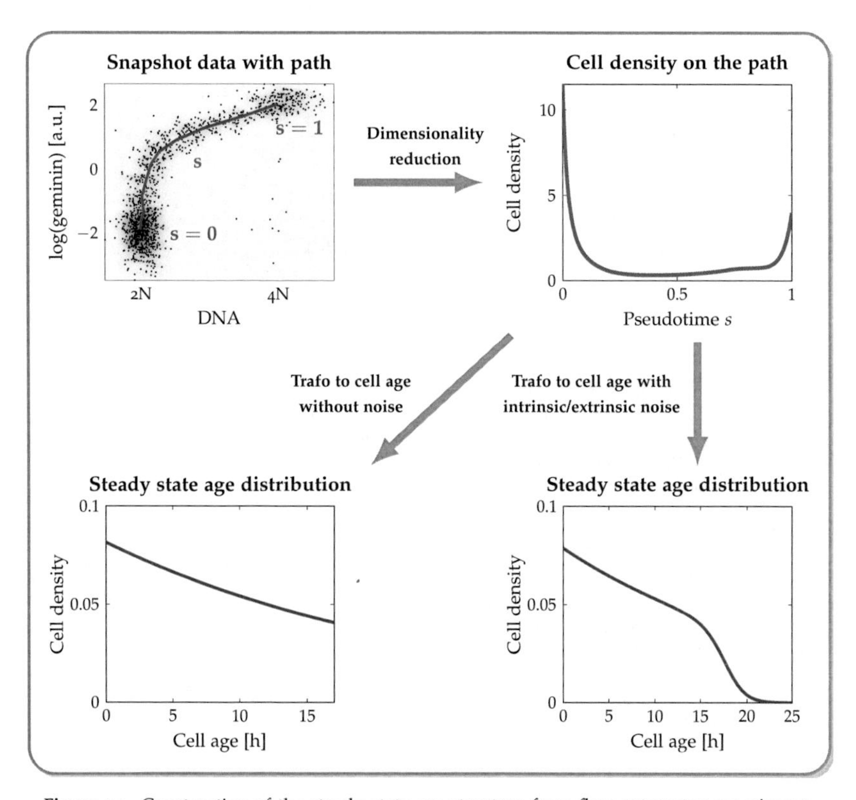

Figure 3.1. Construction of the steady state age-structure from flow cytometry experiments: Levels of DNA (DAPI) and geminin (mAG-hGem) in an exponential growing population of NCI-H460/geminin cells. An average cell follows the indicated path through the dataset. The density of cells along the cell cycle trajectory through the dataset is obtained kernel density estimation on pseudotime values. The cell density in pseudotime is transformed to an age scale resulting in the steady state age-structure of the population without considering noise or with noisy (intrinsic or extrinsic) cell cycle progression.

respectively. Dividing cells cause in addition an influx at $s = 0$ which is twice the outflow at $s = 1$

$$J(0,t) = 2J(1,t). \tag{3.12}$$

Furthermore, the population is in an exponential growth phase implying a linear ordinary differential equation for the change of the total cell number

$$\frac{\mathrm{d}}{\mathrm{d}t}N(t) = \gamma N(t). \tag{3.13}$$

The number density $n(s,t)$ is connected to the population growth via

$$\frac{\mathrm{d}}{\mathrm{d}t}N(t) = \frac{\mathrm{d}}{\mathrm{d}t}\int_0^1 n(\sigma,t)\,\mathrm{d}\sigma = \gamma \int_0^1 n(\sigma,t)\,\mathrm{d}\sigma$$

which is in a stationary case also true for parts of the cell cycle

$$\frac{\mathrm{d}}{\mathrm{d}t}\int_0^s n(\sigma,t)\,\mathrm{d}\sigma = \gamma \int_0^s n(\sigma,t)\,\mathrm{d}\sigma. \tag{3.14}$$

For the transformation of a measured cell number density along the path s (as shown in Figure 3.1) to a cell number density with respect to age a, the so far unknown functions $f(s)$ and $g(s)$ of the original model Eq. (3.7) must be identified. An expression for the velocity function $f(s)$ in the case without noise ($g(s) = 0$) is given by

$$f(s) = \gamma\frac{2 - P(s)}{p_s(s)}. \tag{3.15}$$

This function is only dependent on quantities that are known from experiments, namely the population growth rate γ, the probability distribution $p_s(s)$ and the cumulative distribution $P(s)$ of cells on the path s. The expression is identical to the one stated in Kafri et al. 2013. A detailed derivation of Eq. (3.15) is given in Appendix E.1.

3.3.2 Transformation to steady state age-structure without noise

The transformation of the cell number density in pseudotime $p_s(s)$ to a cell density with respect to age a is achieved in a few steps. First the transformation $\tau : s \mapsto a$ is derived which is then applied to the measured cell number density by using the law for the transformation of probability densities. In the case without noise ($g(s) = 0$), the SDE Eq. (3.7) for the progression of a single cell through the cell cycle reduces to a simple ODE

$$\frac{\mathrm{d}s}{\mathrm{d}t} = f(s) \tag{3.16}$$

which can be solved by separating the variables and subsequently integrating both sides. Substituting $f(s)$ with Eq.(3.15) results then in the function $\tau(s) = a$ which relates the pseudotime value s to a cell age a:

$$\tau(s) = \frac{1}{\gamma}\ln\left(\frac{2}{2 - P(s)}\right). \tag{3.17}$$

For the transformation of $p_s(s)$ to cell density with respect to its age $p_a(a)$, the transformation rule for probability functions must be used to preserve the area under the curve

$$p_a(a) = \left| \frac{\mathrm{d}}{\mathrm{d}\,a} \tau^{-1}(a) \right| \; p_s\left(\tau^{-1}(a)\right). \tag{3.18}$$

Analytic expressions can be obtained for the inverse transformation and its derivative

$$\tau^{-1}(a) = P^{-1}\left(2 - 2e^{-a\,\gamma}\right) \tag{3.19}$$

$$\frac{\mathrm{d}}{\mathrm{d}\,a} \tau^{-1}(a) = 2\gamma e^{-a\,\gamma} \, \frac{1}{p_s\left(\tau^{-1}(a)\right)}. \tag{3.20}$$

By substituting Eq. (3.20) in Eq. (3.18) the measured cell number density cancels out. Hence, the remaining cell number density with respect to age is identical to the steady state age-structure Eq. (3.6) obtained from age-structured population models

$$p_a(a) = \left| 2\gamma e^{-a\,\gamma} \, \frac{1}{p_s\left(\tau^{-1}(a)\right)} \right| \; p_s\left(\tau^{-1}(a)\right) = 2\gamma e^{-a\,\gamma}. \tag{3.21}$$

This demonstrates the correctness of the derived transformation. For the application to measured data the transformation can be easily applied to cell number densities in pseudotime and measured outputs $p(s, \mathbf{y})$ to obtain the age and output dependent cell number density of the population

$$p_a(a, \mathbf{y}) = \left| 2\gamma e^{-a\,\gamma} \, \frac{1}{p_s\left(\tau^{-1}(a)\right)} \right| \; p_s\left(\tau^{-1}(a), \mathbf{y}\right). \tag{3.22}$$

The first column in Figure 3.4 illustrates the application of the method to a dataset from flow cytometry experiments in a NCI-H460/geminin cell population where the *geminin* and *DAPI* signals were used for the path construction.

3.3.3 Transformation to steady state age-structure with stochastic progression rate

As stated in the model formulation in Eq. (3.7), cell cycle progression is influenced by noise and thus there is no one-to-one relation between pseudotime s and the age a of a cell. A transformation from pseudotime to the age scale can, however, directly be stated from our model formulation with the time evolution of the number density of a cell population in pseudotime, given by the PDE in Eq. (3.8): The solution of the PDE, when initialized with a Dirac delta distribution at the origin $n_0(s) = \delta(s)$ provides a probability measure for a cell to have pseudotime s, given that it has age a

$$n(s, t) \to p(s|a = t). \tag{3.23}$$

This probability measure provides the transformation for a single generation of cells, thus the influx in Eq. (3.12) is omitted ($J(0, t) = 0$) and the reflecting and absorbing boundary Eqs. (3.10) and (3.11) are used. The density in pseudotime $p_s(s)$ can then be transformed to the number density of cells with respect to age by the convolution

$$p_a(a) = \int_0^1 p(\sigma|t = a) p_s(\sigma) \,\mathrm{d}\,\sigma. \tag{3.24}$$

Similar as in the simplified case without noise, this can again be applied to the density in pseudotime and measurement outputs $p(s, \mathbf{y})$ to calculate the desired number density of the population with respect to age a and the additional measured quantities $\mathbf{y}$

$$p_a(a, \mathbf{y}) = \int_0^1 p(\sigma | t = a) p_s(\sigma, \mathbf{y}) \, \mathrm{d}\sigma. \tag{3.25}$$

Solving the PDE Eq. (3.8) requires the functions $f(s)$ and $g(s)$ which are not known apriori. In line with the approach to derive $f(s)$ in the case where the noise is omitted (Eq. (3.15)), one obtains when considering the noise, an expression for $f(s)$ which contains the unknown noise term $g(s)$

$$f(s) = \gamma \frac{2 - P(s)}{p_s(s)} + \frac{1}{2} \frac{\frac{\partial}{\partial s} \left(g(s)^2 p_s(s) \right)}{p_s(s)}. \tag{3.26}$$

Regarding the SDE model of single cell progression through the cell cycle, it is important to note that the distribution does not change with time. Furthermore, the relation between the densities at given positions are conserved over different parameterisation of the path. As a result, one can define a reference curve parameterisation for which the speed and the noise terms are known, or can be identified. Such an arbitrary curve parameterisation x has in its most general form a SDE for each cell with cell speed $v(x)$ and the noise function $h(x)$

$$\mathrm{d}x = v(x) \, \mathrm{d}t + h(x) \, \mathrm{d}W. \tag{3.27}$$

The corresponding Fokker-Planck equation for the evolution of the cell density is given by

$$\frac{\partial n(x,t)}{\partial t} = -\frac{\partial}{\partial x} \left(v(x) n(x,t) \right) + \frac{1}{2} \frac{\partial^2}{\partial x^2} \left(h(x)^2 n(x,t) \right). \tag{3.28}$$

The noise function $g(s)$ for the s-parameterisation can be calculated with a function $\varphi : s \mapsto x$, which transforms the SDE in s Eq. (3.7) to the reference SDE in x Eq. (3.27) for which the speed and velocity functions are known. The detailed procedure is given in E.2. In the simplest case speed and noise are just constants, v and D which can be identified, for example, from the distribution of total cell cycle lengths in the population (described in Section B.6). If $g(s)$ is identified, $f(s)$ is obtained from Eq. (3.26) and the PDE Eq. (3.8) with reflecting and absorbing boundary Eqs. (3.10) and (3.11) can be solved numerically. The second column in Figure 3.4 illustrates the application of the method to the same dataset as described in Section 3.3.2.

The cell number density with respect to age must, when applying the transformation Eq. (3.24), be identical to the steady state age-structure Eq. (3.5) that one obtains from age-structured population models. The age-dependent division rate $\alpha(a)$ in Eq. (3.5) can be obtained from the distribution of the total cell cycle lengths in a population, which must be equal to the steady state age-structure multiplied with the division rate (see E.3 for details). A comparison of the transformed age-structure with the theoretical age-structure of age-structured population models Eq. (3.5) illustrates the correctness of the derived transformation (Figure 3.2 a).

3.3.4 Transformation to steady state age-structure with distributed progression rate

Persistent variance in cell cycle progression is described by a distributed cell cycle progression rate $f(s)$ without any stochastic contribution ($g(s) = 0$). This rate can directly be conferred from the distribution of total cell cycle lengths that may be observed in live-cell microscopy experiments. Given the direct relation between the cell cycle length T and the growth rate γ of a population

$$\gamma = \frac{\ln 2}{T} \tag{3.29}$$

a distributed cell cycle length in a population with division probability $p_c(T)$ results in a distributed transformation $\tau(s|T)$. The equivalent transformation as in the case without noise (Eq.(3.18)) is then given by a conditional probability for observing a cell with age a given a specific growth rate or total cell cycle length

$$p_a(a \mid T) = \left| \frac{\mathrm{d}}{\mathrm{d}a} \tau^{-1}(a \mid T) \right| p_s\left(\tau^{-1}(a \mid T)\right). \tag{3.30}$$

The actual age-dependent steady state distribution is obtained by marginalizing the conditional probability over the distributed total cell cycle length

$$p_a(a) = \int_0^\infty p_c(T)\, p_a(a \mid T)\, \mathrm{d}T \tag{3.31}$$

Again, this transformation can be applied to a cell number density in pseudotime and measurement outputs $p(s, \mathbf{y})$ to get the desired number density of the population with respect to age a and the measured quantities

$$p_a(a, \mathbf{y}) = \int_0^\infty p_c(T)\, p_a(a, \mathbf{y} \mid T)\, \mathrm{d}T. \tag{3.32}$$

The cell number density with respect to age derived from Eq. (3.31) must be identical to the steady state age-structure Eq. (3.5) obtained from age-structured population models. A comparison of the transformed age-structure with the theoretical age-structure of age-structured population models Eq. (3.5) illustrates the correctness of the derived transformation (Figure 3.2 b).

3.4 Cell cycle-dependent expression of Cyclin B1

We now asked the question, which transformation method can capture the variance in cell age observed in populations. We therefore applied the transformation methods (without noise, with intrinsic and extrinsic noise) to a dataset from flow cytometry experiments of asynchronous growing NCI-H460/geminin cells in their exponential growth phase. In such a condition the proportion of cells in each phase of the cell cycle is stable and the age-structure of the population does not change. The cell cycle position of individual cells was determined from DNA content and geminin expression (mAG-hGem(1/110)) (Sakaue-Sawano et al. 2008) (Figure 3.3). mAG-hGem(1/110) (hereafter referred to as geminin) serves as a marker for the

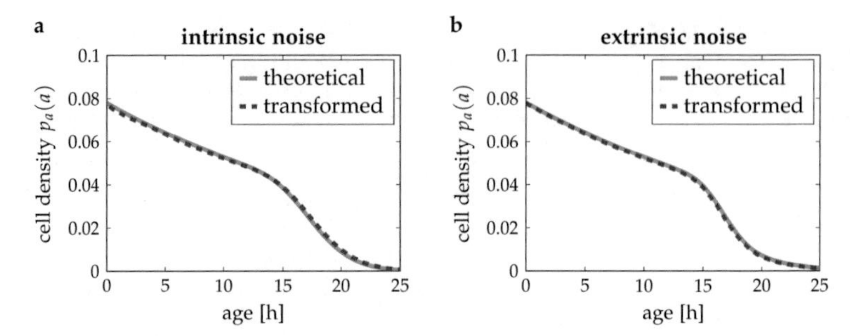

Figure 3.2. Comparison of the transformed age-dependent cell density $p_a(a)$ with intrinsic and extrinsic noise with the theoretical steady state distribution from age-structured population models. (**a**) The intrinsic noise in cell cycle progression described by a SDE was estimated from the distribution of total cell cycle lengths. (**b**) Extrinsic noise in the population is captured by a distributed cell cycle progression rate which is directly obtained from the distribution of total cell cycle lengths.

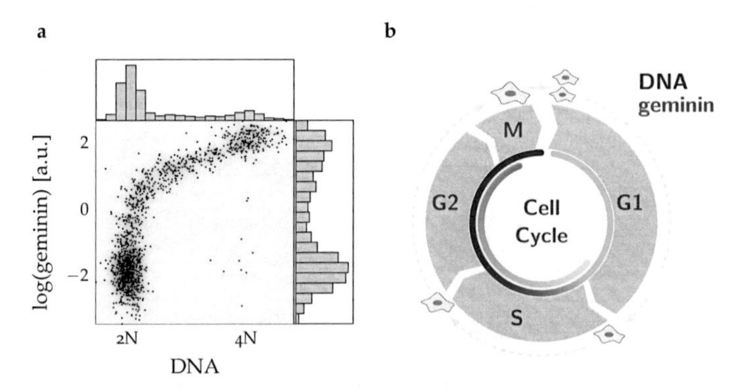

Figure 3.3. (**a**) Levels of DNA (DAPI) and geminin (Cyclin B1 not shown) in an unsynchronised, exponential population of NCI-H460/geminin cell obtained from flow cytometry experiment analysis. (**b**) Schematic drawing of mAG-hGem(1/110) expression used for cell cycle progression path construction.

activity of the anaphase-promoting complex (APC) in its active APC^{Cdh1} form. The mean cell cycle trajectory path in Figure 3.3 illustrates the relation between the cell cycle stage and the position in the DNA vs. geminin plane. Expression of Cyclin B1 was analyzed in parallel and not used for the construction of the cell cycle path, but serves as a verification for the method as cell cycle-dependence of Cyclin B1 is well known (Morgan 2007; Pomerening et al. 2003; Santos et al. 2012). The resulting change of the number density of cells with respect to their age and the Cyclin B1 abundance exhibits the characteristic two step process (Figure 3.4 first row). An initial rise in Cyclin B1 is seen for cells with an age of 6-8 hours, which is slightly before the entry into S phase, as can be seen from a comparison with the DAPI signals (Figure 3.4 second row). The Cyclin B1 expression hardly changes during S phase. However, a strong increase in its expression is observed during G2/M phase which is in accordance with its biological function as mediator of the G2/M phase transition. This behavior is seen in all transformed datasets, though the transitions in the different phases are more blurred if the noise is considered during the transformation.

3.4.1 Age distribution is preserved in transformation with noise

In the following, we compare the results from the transformations to age distributions captured in time-lapse microscopy experiments. To this end, individual cells were tracked and the time it took for the cells from their last division to the transition from G1 to S phase was determined. The G1-S transition is characterized in live-cell microscopy experiments with NCI-H460/geminin cells by the visible expression of the geminin reporter.

The corresponding geminin expression in the flow cytometry dataset is indicated by a red line in Figure 3.5 a. A comparison of the age distributions at the G1-S transition obtained from live cell microscopy and the one obtained from the transformations is shown in Figure 3.5 b. The distribution of cells at the G1-S transition is quite narrow in the case where the noise was not considered during the transformation. One reason for the narrow distribution is that the cell number density in pseudotime $p_s(s)$ is low in the relative large range of s where geminin has its strongest increase (see Figure 3.1). A low density implies a high progression rate $f(s)$ which in turn results in a transformation to a narrow age range.

Noise consideration in the transformation method results in a wider spread of the age distribution at G1-S transition which is closer to the distribution of the samples. In case of intrinsic noise this wider spread is caused by the spread in the probability measure $p(s \mid a = t)$ which was used for the transformation. The probability measure was obtained by solving the PDE Eq. (3.8) for the evolution of the number density of cells in pseudotime in which the noise in the population was taken into consideration by adding a Wiener process. Extrinsic noise causes a wider spread due to the distributed cell cycle progression rate which results in a distributed transformation $\tau(s \mid T)$.

While the transformation with intrinsic noise overestimates the G1-S spread the transformation with extrinsic noise underestimates it. The negative logarithm of the likelihood (nLL) to observe the measured G1-S transition times with the distributions obtained from both transformations with noise is very similar and significantly better than the one neglecting noise. Thus, by accounting for noise in the transformation the age distribution at the G1-S transition is very similar to the one obtained from imaging experiments while the approach without noise exhibits a too narrow age distribution for this cell cycle related event (Figure 3.5 b).

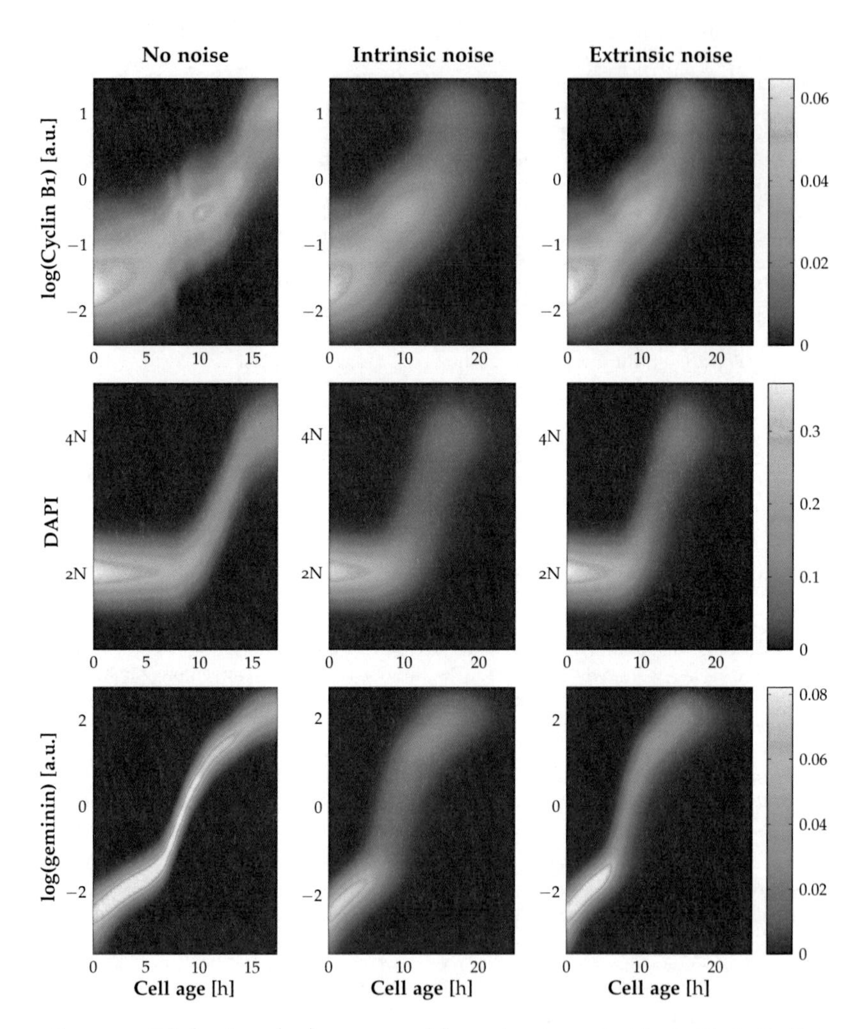

Figure 3.4. Cell density $p_a(a, \mathbf{y})$ over age and flow cytometry readout levels (Cyclin B1 (first row), DNA (second row) and geminin (third row)) of exponentially growing NCI-N460 cells. Figures in the first column were obtained from the transformation without noise. The second and third column illustrate the results from the transformations where intrinsic or extrinsic noise in cell cycle progression is taken into account. The dynamical behavior of the different cell cycle related readouts is in line with the expected trajectory. The distributions in the results from transformation with noise are more blurred than those without noise, thereby reflecting the dispersing age distribution through the cell cycle.

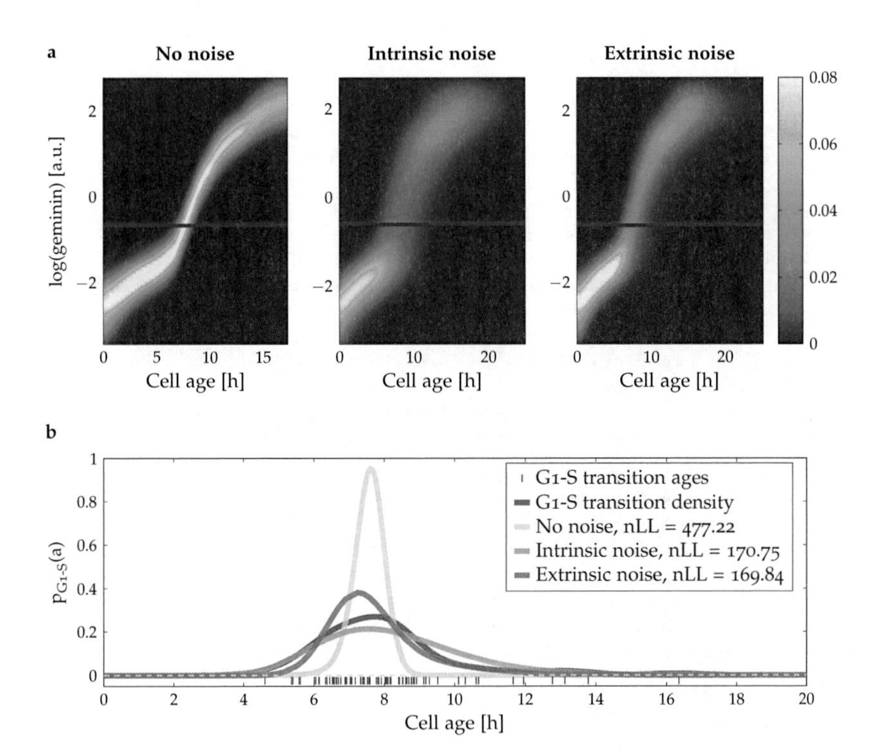

Figure 3.5. Comparison of G1-S transition age distributions $p_{G1-S}(a)$ obtained from EA transformations and live-cell microscopy. The G1-S transition is characterized in live-cell microscopy experiments with NCI-H460/geminin cells by the visible expression of the geminin reporter. (**a**) The corresponding geminin expression in the transformed flow cytometry dataset without noise and with intrinsic and extrinsic noise consideration is indicated by a red line. (**b**) Age distributions from EA transformations and live-cell microscopy.

3.5 Summary and discussion

Summary

In this chapter, we presented the relationship between cell cycle analysis methods based on ergodic principles and age-structured population models. Based on a description of the progression of a single cell through the cell cycle by a stochastic differential equation, we derived transformation methods that relate pseudotime to cell age. In the simple case where the noise was omitted, we showed analytically, that the derived transformation results in the steady state age-structure of age-structured population models. By establishing this connection, the rich theory on age-structured population models can be transferred to cell cycle analysis methods based on ergodic principles (Cushing 1998). We furthermore presented two methods that preserve the noise in the population during the transformation by accounting either for intrinsic or extrinsic noise. In the case of intrinsic noise, the SDE model of cell progression in pseudotime is identified with the help of another SDE of which speed and noise functions are known. This SDE can be in an arbitrary parameterisation, which implies minimal requirements for the known progression and noise functions. We chose in our example the simplest parameterisation where progression and noise functions are constants which we identified from the cell cycle length distribution in a population. Additional information about the fluctuations in cell cycle progression (e.g. from live-cell time-lapse microscopy or synchronization experiments) can be used to deduce a more accurate noise description in relation to cell cycle progression. This would reduce the uncertainty in the transformation and improve the accuracy of the transformed distributions.

Extrinsic noise is characterized by a distributed cell cycle progression rate which can be deduced from the distribution of total cell cycle lengths in the population. The presented method is restricted to a constant variance in the cell cycle progression. This restriction can be relaxed by partitioning the cell cycle in phases with different cell cycle progression rate distributions. Each phase must then be transformed to the corresponding age-structure of an age-structured population model divided into the same phases (e.g. Billy et al. (2011)). The level of variation in the different phases is then related to the age-dependent transition rate from one phase to the next.

Discussion and Outlook

Our methods are based on two assumptions: (1) the population is in its steady state distribution and (2) variance in the cell cycle progression is either introduced by a stochastic process or a distributed rate. The first assumption must be verified experimentally and is fundamental for methods based on ergodic principles. The established relation with age-structured population models, however, provides the theoretical basis for the analysis of populations that deviate from their steady state in the course of an experiment. Deviations from the steady state distribution due to cell cycle arrest or cell death can be captured in the form of a change of the cell cycle progression rate or a cell death rate in structured population models. One could thereby retain the relation between the path and cell age and thus infer cell cycle-dependent rates of molecular events also in populations that are not in their steady state age distribution. We present these extensions in Chapter 4.

The description of cell cycle progression by a single cell with a stochastic differential equation implies the rather strong assumption that all cells in the population proceed through the cell cycle with an identical progression rate $f(s)$ from which deviations occur randomly due to the Wiener process. We did not observe any correlation between cell cycle lengths of descendent cells in live-cell experiments. Furthermore the length of G1 phase and the proceeding cell cycle phases (S, G2 and M) was also not correlated, thus supporting a random nature for deviations in cell cycle progression. However, Sandler et al. (2015) observed strong correlations between cell cycle times of sister cells in time-lapse images of dividing L1210 Fucci cells, indicating a more extrinsic noise type. The true sources of variability in cell cycle progression are still unknown and probably compromise both intrinsic and extrinsic noise. The transformations presented here could in principle be combined to account for stochastic and distributed cell cycle progression.

The application to a DAPI, geminin and Cyclin B1 dataset from flow cytometry experiments provided insights in the capabilities of the methods. The transformation without noise results in a one-to-one relation of pseudotime and the cellular age. Although, this is in general not the case, the local change of the measured quantity recovers the rate of change at the specific cell cycle position. This fact can be seen in the characteristic linear increase of DNA during the S-phase (Figure 3.4 c). Noise consideration in the transformation leads to the representation of each position on the path by a distribution on the age scale. The resulting distribution is in turn more disperse, which does not mean that the method is less accurate. The method rather retains the true age distribution of cells in the population at the different cell cycle stages as shown for the age distribution at the G1-S transition in Figure 3.5.

Taken together, noise consideration causes a transformation to a true age scale while the transformation without noise results in a transformation to a cell cycle stage scale which locally equals time. The presented connection of the EA methods to age-structured population models is in general applicable (with slight adjustments) to any process that exhibits a steady state distribution ranging from cell cycle related processes in bacteria or budding yeast to stem cell development. Furthermore, the combination of the EA methods with age-structured population models can facilitate the application of the method to non-steady state distributions, which were previously excluded as the steady state assumptions was central to the method.

Chapter 4

Cell cycle progression inference

This chapter is based on the manuscript:

> Karsten Kuritz et al. (2020a). 'PDE-constrained optimization for estimating population dynamics over cell cycle from static single cell measurements'. In: *bioRxiv*. DOI: 10.1101/2020.03.30.015909.

Processes in biological systems can nicely be studied while being in steady state as we have shown in the previous chapters. However, analysing a system in response to some perturbation such as environmental changes or drug treatment is of major importance to fully characterize the system behavior. Hence, a compelling extension of the previous approaches are studies where the stationary assumption is not satisfied. This chapter presents an extension of MAPiT to non-stationary processes. The chapter deals with the specific example where we want to infer altered cell cycle progression in response to treatments as illustrated in Figure 4.1. By minimizing the difference between cell cycle-dependent number densities obtain with MAPiT and the solution of a parameterized age-structured population model we estimate changes in cell cycle progression in response to treatments. Our newly derived sensitivity system furthermore enables efficient solving of the problem. We showcase our framework by estimating the changes in cell cycle progression in K562 cells treated with Nocodazole and identify an arrest in M-phase transition that matches the expected behavior of microtubule polymerization inhibition. Our results have two major implications: First, identifying changes in cell cycle progression in response to treatments provides scope to characterize the effect on cell cycle progression of known and unknown compounds in large-scale screenings. Second, knowledge of the cell cycle stage- and time-dependent progression function facilitates a transformation from pseudotime to real-time thereby enabling real-time analysis of molecular rates in response to treatments. In this way, our method contributes significantly towards the construction of quantitative whole cell models.

The experimental data that we present in this chapter was prepared by the El-Samad Lab at the University of California, San Francisco. This chapter is taken in parts from Kuritz et al. (2020a).

4.1 Background and problem formulation

Cells are regularly exposed to various stresses, including environmental changes or drug treatment. One of the first cell responses to environmental stresses is to modulate its cell cycle, usually stopping its progression. Cell cycle arrest occurs at any of the four cell cycle

stages (G1, S, G2, M), and it is dynamically regulated (Kapuy et al. 2009).
Identifying cell cycle progression changes caused by environmental factors, or treatments, is crucial for the understanding of cellular responses and the development of new treatment options for many pathologies. However, measuring changes in cell cycle progression is experimentally challenging and model inference computationally expensive. Several algorithms designed to reconstruct cell trajectories from single-cell data, including Wanderlust, Monocle and diffusion maps (DPT), have been recently proposed (Saelens et al. 2019; Bendall et al. 2014; Fabian J. Theis et al. 2015; Qiu et al. 2017). These methods order single-cell data in pseudotime – a quantitative measure of progress through a biological process. In Chapter 2, we introduce MAPiT, a computational framework that transforms pseudotime into real-time. Here, we apply MAPiT to detect dynamically-regulated changes in cell cycle progression by learning a time- and cell cycle position-dependent function from experimentally observed cell distributions (Figure 4.1). We introduce a computational framework that allows the inference of changes in cell cycle progression from static single-cell measurements thereby solving the following problem:

Problem 4.1. Given time-course of single-cell data, identify the time- and cell cycle position-dependent changes in cell cycle progression.

Over the course of this chapter, we will first introduce the experimental data and the way we use the methods from Chapter 2 for data processing. Then, we present the partial differential equation (PDE) to model cell cycle-dependent cell density and the cost function that form the PDE-constrained optimization problem. Our newly derived sensitivity system enables efficient solving of the problem. Finally, we demonstrate the capability of the algorithm with one artificial and two experimental datasets.

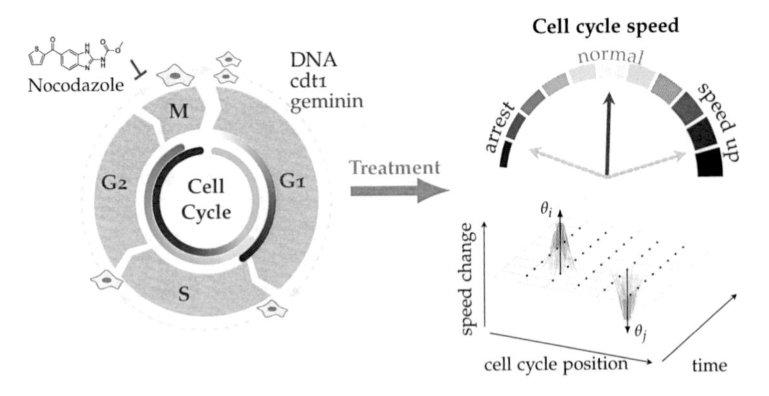

Figure 4.1. Schematic cell cycle progression with markers (DNA, cdt1, geminin) and Nocodazole treatment. A treatment causes time- and cell cycle position-dependent change in the cell cycle speed.

4.2 Cell cycle progression inference with sensitivities

In this section, we present the main results of this chapter. First, we present data preprocessing steps, followed by the derivation of the PDE-constrained optimization problem for the estimation of changes in cell cycle progression.

4.2.1 Preprocessing of experimental single-cell data

To enable the identification of changes in cell cycle progression, we preprocessed the static data from time series single-cell experiments (Figure 4.2). In summary, we (1) found the path through the dataspace using pseudotime algorithms; (2) derived cell density on the path; (3) transformed to cell cycle position-dependent cell density with MAPiT; and (4) calculated the relative cell number density.

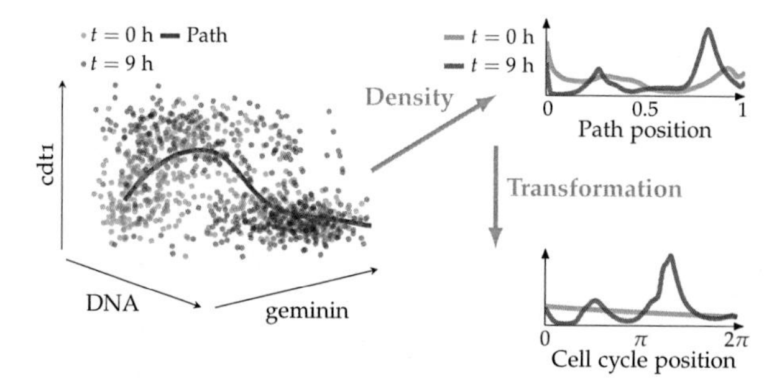

Figure 4.2. Data preprocessing steps shown for 0 and 9 h treatment with Nocodazole: Pseudotime algorithms find a path in the dataspace which an average cells takes during cell cycle progression. Kernel density estimation followed by dimensionality reduction provides a cell density on the path. Applying a transformation learned from unperturbed populations results in cell densities on a cell cycle position scale.

To find the path in the dataset at $t = 0$ h, we applied a previously described pseudotime algorithm (Bendall et al. 2014). We probed different root cells for the pseudotime axis, and created the average path that a cell takes through the dataspace $l : [0, 1] \to \mathbb{R}^3$ by a moving average of signal intensities along the pseudotime axis. To obtain the cell density $p_s(s)$ along the path, we constructed multi-dimensional Gaussian distributions centered on the data coordinates μ_i of each cell

$$p_y^i(\mathbf{y}) = \mathcal{N}(y \mid \mu_i, \Sigma) \; . \tag{4.1}$$

Bandwidths Σ were set with Silverman's rule of thumb (Silverman 1986). We evaluated these distributions for each cell on the path $l(s)$ and normalized resulting densities to one

$$p_s^i(s) = p_y^i(l(s)) \left(\int_0^1 p_y^i(l(\sigma)) \, \mathrm{d}\sigma \right)^{-1} \; . \tag{4.2}$$

The cell number density along the path was given by the sum of all individual probability densities normalized by the number of cells

$$p_s(s) = \frac{1}{N}\sum_{i=1}^{N} p_s^i(s) . \tag{4.3}$$

The density on the path $p_s(s)$ depends on various factors, such as, the measured markers or the cell type. A transformation from the path position scale to an independent cell cycle position scale (normalized time scale) solves this issue. Applying MAPiT, we derived this transformation $\tau : s \rightarrow x$ with $x \in [0, 2\pi]$. The densities at $t = 0$, $p_s^0(s)$, were transformed to the steady-state distribution of an unperturbed population of cycling cells (Powell 1956; Kafri et al. 2013; Kuritz et al. 2017)

$$\bar{p}_x(x) = \frac{ln(2)}{\pi} 2^{-\frac{x}{2\pi}} . \tag{4.4}$$

The cumulative density functions of the distributions were used to obtain the transformation.

$$\begin{aligned} \tau^{-1}(x) &= \left(P_s^0\right)^{-1}\left(\bar{P}_x(x)\right) , \\ \text{or} \quad \tau(s) &= \left(\bar{P}_x\right)^{-1}\left(P_s^0(s)\right) . \end{aligned} \tag{4.5}$$

Next, we calculate the distribution of cells in the cell cycle by applying this transformation to the density on the path at every time point

$$p_x^k(x) = \left|\frac{\mathrm{d}\,\tau^{-1}(x)}{\mathrm{d}\,x}\right| p_s^k\left(\tau^{-1}(x)\right) . \tag{4.6}$$

The procedure is illustrated for two time points in Figure 4.2 for a three-dimensional data-space. Finally, we derived the cell number density $\eta_k(x)$ at time points t_k by multiplication of the densities $p_x^k(x)$ with the respective population size N_k:

$$\eta_k(x) = N_k\, p_x^k(x) . \tag{4.7}$$

4.2.2 Model and optimization problem

We then asked what are the time- and cell cycle position-dependent changes in cell cycle progression speed, that reproduce experimentally observed time-course of cell number densities. To find the speed change function, we ran an optimization protocol that minimized the difference between the cell number densities from experimental data $\eta_k(x)$ at time-points t_k and the model predictions $n(x, t_k)$. The optimization was constrained by requiring that the predicted number density is a solution of the age-structured population model type PDE

$$\begin{aligned} \frac{\partial n(x,t)}{\partial t} + \gamma \frac{\partial (v(x,t|\theta)\, n(x,t))}{\partial x} &= 0 , \\ n(x,0) &= \bar{p}_x(x) , \\ n(0,t) &= 2\, n(2\pi, t) \end{aligned} \tag{4.8}$$

with $\gamma = \frac{T}{2\pi}$, where T is the period of the cell cycle.

The speed change function $v(x, t|\theta)$ is such that $v(0, x|\theta) = 1$ which resulted in the classical age-structured population model for $t = 0$ h (Kuritz et al. 2017; Foerster 1959). Parameters

$\theta \in \mathbb{R}^m$ are the weights of m 2-D normal distributions centered at grid points $(\mu_i)_{i=1,\dots,N}$ in the t and x domains. To ensure the condition for $t = 0$ h, we defined the speed change function as the exponential of a sum of Gaussians multiplied with a hill function in t:

$$v(x,t|\theta) = \exp\left(\frac{t^5}{\kappa^5 + t^5}\sum_{i=1}^{N}\theta_i \mathcal{N}(x,t \mid \mu_i, \Sigma)\right) . \tag{4.9}$$

The bandwidth was chosen manually, depending on the number and spacing of grid points. An example of a speed change function is shown in Figure 4.1.

We used a symmetric Kullback-Leibler divergence ($\mathcal{KL}_s$) to quantify the deviation of model predictions from experimental data. The objective function for the parameter estimation was set as the sum of the logarithm of the $\mathcal{KL}_s$ at all time points. Suppose, $p(x)$ and $q(x)$ are two probability distributions; then the symmetric Kullback-Leibler divergence is given by

$$\mathcal{KL}_s(p \parallel q) = \int_{\mathcal{X}} (p(x) - q(x)) \log\left(\frac{p(x)}{q(x)}\right) \mathrm{d}x , \tag{4.10}$$

then the objective function reads

$$F(\theta) = \sum_{k=1}^{K} \log(\mathcal{KL}_s(\eta_k \parallel n(\cdot, t_k|\theta))) . \tag{4.11}$$

To minimize the difference between data $\eta_k(x)$ and model prediction $n(x, t_k|\theta)$, we pose the following optimization problem:

$$\begin{aligned} &\min_{\theta} F(\theta) \\ &\text{s.t.} \quad n(x, t_k|\theta) \text{ is a solution of Eq. (4.8)} \end{aligned} \tag{4.12}$$

This PDE-constrained optimization problem is non-convex and possesses several local minima.

4.2.3 Sensitivity system for efficient parameter estimation

The availability of gradients for the objective function concerning the parameters dramatically improves the solution of optimization problems. These gradients can be approximated by finite difference, or computed as the solution of a sensitivity system. Derivative-based optimization employing the sensitivity equations is known to outperform other methods by orders of magnitude in accuracy and speed (Raue et al. 2013). To enable the efficient parameter estimation for the PDE-constrained optimization problem, we derived the gradient of the objective function as

$$\frac{\partial F(\theta)}{\partial \theta} = \sum_{k=1}^{K} \frac{\int_{\mathcal{X}} S(x,t_k|\theta)\left(1 - \frac{\eta_k(x)}{n(x,t_k|\theta)} - \log\frac{\eta_k(x)}{n(x,t_k|\theta)}\right)\mathrm{d}x}{\mathcal{KL}_s(\eta_k \,||\, n(\cdot,t_k|\theta))} , \tag{4.13}$$

with $S(x,t_k|\theta) = \frac{\partial n(x,t_k|\theta)}{\partial \theta}$ denoting the sensitivity of the predicted number density at time point t_k for parameters θ.

To increase clarity in the derivation of the sensitivity system, we use a shorthand notation where we omit the arguments in the functions S, n, v. Differentiating the system in Eq. (4.8) with respect to the parameters θ provides a PDE for the sensitivities

$$\frac{\partial}{\partial\theta}\left(\frac{\partial n}{\partial t} + \gamma\frac{\partial(vn)}{\partial x}\right) = 0$$
$$\frac{1}{\gamma}\frac{\partial S}{\partial t} + v\frac{\partial S}{\partial x} = -\frac{\partial^2 v}{\partial x\partial\theta}n - \frac{\partial v}{\partial\theta}\frac{\partial n}{\partial x} - \frac{\partial v}{\partial x}S\,. \tag{4.14}$$

In classical ODE-constrained optimization problems, the sensitivities are obtained by solving a system of differential equations (Raue et al. 2013). Unlike in ODE sensitivity systems, closure of the differential equations was not achieved because the unknown function $n_x := \frac{\partial n}{\partial x}$ emerged. Differentiating the system in Eq. (4.8) with respect to x results in a new PDE:

$$\frac{\partial}{\partial x}\left(\frac{\partial n}{\partial t} + \gamma\frac{\partial(vn)}{\partial x}\right) = 0$$
$$\frac{1}{\gamma}\frac{\partial n_x}{\partial t} + v\frac{\partial n_x}{\partial x} = -\frac{\partial^2 v}{\partial x^2}n - 2\frac{\partial v}{\partial x}n_x\,. \tag{4.15}$$

The system of coupled first order nonhomogenous PDEs composed of Eqs. (4.8), (4.14) and (4.15) then constitutes the sensitivity system. Its solution which we calculated numerically provided the sensitivities which we applied in Eq. (4.13) to compute the gradient of the objective function. Taken together, the resulting system of PDEs is given by:

$$\begin{aligned}
\frac{1}{\gamma}\frac{\partial n}{\partial t} + v\frac{\partial n}{\partial x} &= -\frac{\partial v}{\partial x}n \\
\frac{1}{\gamma}\frac{\partial n_x}{\partial t} + v\frac{\partial n_x}{\partial x} &= -\frac{\partial^2 v}{\partial x^2}n - 2\frac{\partial v}{\partial x}n_x \\
\frac{1}{\gamma}\frac{\partial S}{\partial t} + v\frac{\partial S}{\partial x} &= -\frac{\partial^2 v}{\partial x\partial\theta}n - \frac{\partial v}{\partial\theta}n_x - \frac{\partial v}{\partial x}S
\end{aligned} \tag{4.16}$$

With boundary and initial conditions

$$\begin{aligned}
n(x,0) &= \bar{p}_x(x), & n(0,t) &= 2\,n(2\pi,t), \\
n_x(x,0) &= \frac{\partial\bar{p}_x(x)}{\partial x}, & n_x(0,t) &= 2\,n_x(2\pi,t), \\
S(x,0) &= 0, & S(0,t) &= 2\,S(2\pi,t)\,.
\end{aligned} \tag{4.17}$$

4.2.4 Properties of the algorithm

Two concepts are critical for the implementation of computational methods. First, they rely on assumptions about the biological system and experimental setup. Second, they must be computationally feasible (in the sense that they do not need a large amount of time to generate results). The following assumptions underlie the algorithm as presented here:

Assumption 4.1. *Cell death and cell cycle arrest cannot be distinguished unambiguously. For example, the same cell densities can result from all cells slowing down or just a few cells dying.*

Assumption 4.2. *Treatment only changes velocity on path, not direction/route of path in dataspace.*

Assumption 4.3. *Response to the stress at specific cell cycle stage is homogeneous.*

We verified assumptions 4.1 and 4.2 by excluding dead cells, using the light scattering characteristics of live cells and by manual inspection of data and path as shown in Figure 4.2.

Solving large PDE systems is computationally expensive. However, by discretizing the solution in x in n points, we transferred the problem to n decoupled ODE systems by the method of characteristics (Evans and Society 1998).

$$\begin{aligned}
\frac{1}{\gamma}\,\dot{x} &= v & x(0) &= x_0\,,\\
\frac{1}{\gamma}\,\dot{n} &= -\frac{\partial v}{\partial x}\,n & n(0) &= \bar{p}_x(x_0)\,,\\
\frac{1}{\gamma}\dot{n}_x &= -\frac{\partial^2 v}{\partial x^2}\,n - 2\frac{\partial v}{\partial x}n_x & n_x(0) &= \frac{\partial \bar{p}_x(x)}{\partial x}\Big|_{x=x_0}\,,\\
\frac{1}{\gamma}\,\dot{S} &= -\frac{\partial^2 v}{\partial x \partial \theta}n - \frac{\partial v}{\partial \theta}\,n_x - \frac{\partial v}{\partial x}S & S(0) &= 0\,,
\end{aligned} \tag{4.18}$$

The linear time varying (LTV) ODE system in Eq. (4.18) had discontinuities at division events where $x(t) = 2\pi$. We used the AMICI toolbox for simulation of the ODE system in Eq. (4.18) as it is optimized for efficiently solving large systems of ODE (Fröhlich et al. 2017). The LTV system Eq. (4.18) has, for a single discretization, $3 + m$ states, with m denoting the number of parameters. Thus, computational complexity increases as a product of the number of parameters, discretization points, and time points $\mathcal{O}(k\,n\,m)$. In the examples with experimental data we have $k = 6, n = 50, m = 64$. A Single evaluation of the objective function in Matlab R2018b on a laptop with Linux, processor Intel Core i5-4300U CPU @ 1.9GHz x 4, 8 GB RAM, took 3.5 or 0.5 seconds, with or without sensitivities, respectively.

4.3 Evaluation with in silico data

We applied our method to recover parameters from an artificial speed change function. We generated a simple speed change function composed of $m = 9$ 2D Gaussian distributions with arbitrarily chosen weights (Figure 4.3). Forward simulation of Eq. (4.8) provided the cell number densities of an in silico experiment. Estimating the parameters resulted in an almost perfect fit of the predictions of the fitted model with the in silico-generated data. Furthermore, estimated parameter values, and the resulting speed change function strongly resembled those underlying the in silico data (Figure 4.3). This example demonstrated the capability of the PDE-constrained estimation approach to correctly estimate an unknown speed change function from a time course experiment with cell number densities.

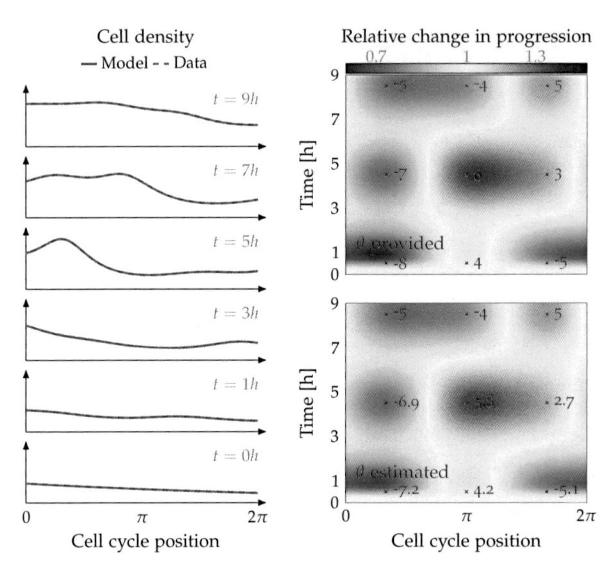

Figure 4.3. Reconstruction of changes in cell cycle progression from artificial data. After parameter estimation, model predictions (solid, red) perfectly recapitulate the cell densities (dashed, blue) generated from the artificial speed change function (top right). Estimated parameters and the resulting function (bottom right) coincide with the true speed change function and the underlying weights (values shown) for the Gaussian distributions.

4.4 Analysis of cell cycle progression

We further tested our method with two proof of concept datasets: (1) control (untreated) and (2) Nocodazole treatment. Nocodazole is an antineoplastic agent that interferes with the polymerization of microtubules, and inhibits the formation of mitotic spindles during M-phase. Therefore, we expected that a slow down/arrest of cell cycle progression would occur at the very end of the cell cycle. In contrast, the control experiment was expected to have constant speed change function ($v(x,t) = 1$). This would result in the exponential growth of the population with a cell cycle distribution equal to the steady-state Eq. (4.4).

After preprocessing the experimental data of the control cells as described above, the calculated distributions deviated significantly from the expected steady-state distribution (Figure 4.4). Accordingly, the estimated speed change function varied strongly depending on time and cell cycle position. In particular, while at later time points, the cell cycle speed in G1 and G2 slowed down, cell cycle progression in the middle of the cell cycle, presumably S-Phase, remained unaltered. We hypothesized that culture conditions during the experiment, such as media exhaustion and increase in cell density, might cause the observed alterations in cell cycle progression.

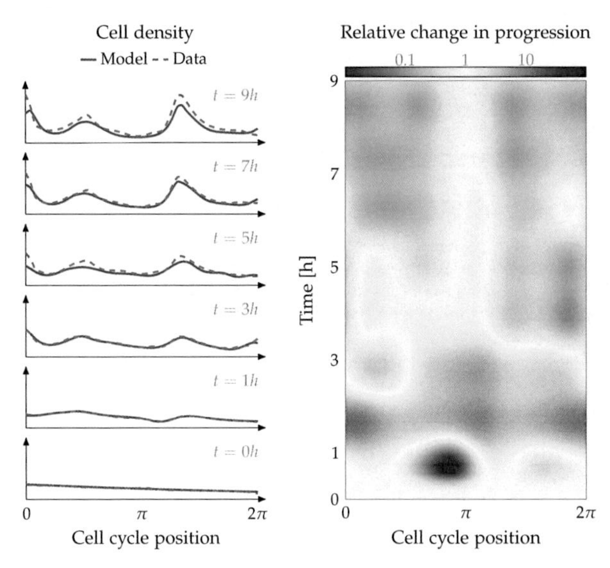

Figure 4.4. Estimated changes in cell cycle progression in a control experiment with untreated cells. Model prediction (solid, red) recapitulates the experimental data (dashed, blue). Two emerging populations are associated with a slow down of cell cycle progression in G1 and G2, but not S phase.

Cell densities in the experiment with Nocodazole treatment showed a similar pattern with accumulating subpopulations at late G1 and the G2-M transition (Figure 4.5). However, the population at the G2-M transition increases significantly throughout the experiment.

This reduced outflow of cells from G2-M was confirmed by the estimated speed change which showed an almost complete arrest at the end of the cell cycle. This behavior is in line with the established mode of action of Nocodazole. Accordingly, and in contrast to the control experiment, the cell density at the beginning of the cell cycle approached zero, further indicating a complete inhibition of cell division.

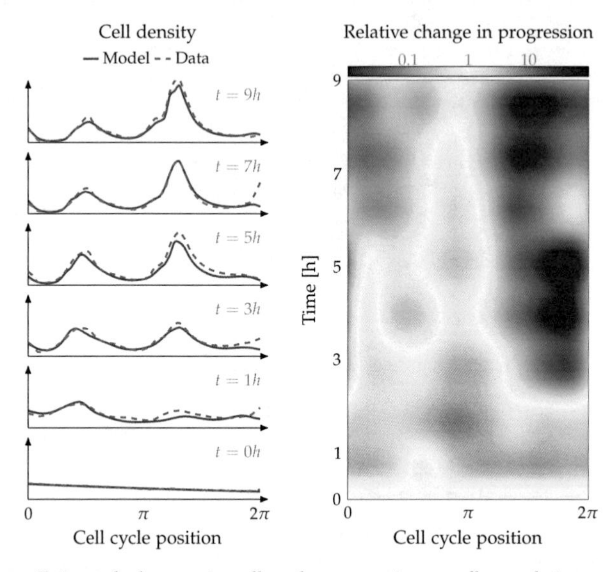

Figure 4.5. Estimated changes in cell cycle progression a cell population treated with $100\,ng/ml$ Nocodazole. Model prediction (solid, red) recapitulates the experimental data (dashed, blue). Accumulation of cells at the end of the cell cycle is caused by complete arrest during M-phase which is inline with Nocodazole mode of action.

4.5 Summary and discussion

Summary

In summary, we presented a computational framework that allows efficient inference of changes in cell cycle progression. To achieve this, we derived the sensitivity system of parameters in the progression function for age-structured type population models. This system allowed the efficient optimization of the objective function, thereby minimizing the difference between experimental observations and model predictions. We showcased the capacity of the method to recover unknown parameters with simulated data. When applied to two experimental datasets, our method uncovered new insights. First, it recognized in the estimated speed change function that Nocodazole induced cell cycle arrest in M-phase. Second, the effects of batch culture cause significant changes in cell cycle progression. Our experimental setup is common for a large class of experiments, and we showed that cell cycle progression is affected by these conditions. Our method can be used to monitor the population for such side effects.

Discussion and Outlook

The presented framework serves as a basis and can be extended or modified to broaden its applicability. A mixture model approach, for example, where two or more populations react distinctly to the same stress could be realized by merely weighting the contribution of these populations to the objective function. In this way, one could overcome assumption 4.3 on the homogeneity of the population's response. While the descriptive speed change function and the resulting molecular rates do not provide mechanistic insights about the cellular response, we envision a substitution of these functions with, e.g. mechanistic ODE models. Such method would require a mapping from the ODE state space to a cell cycle position. The concept of isochrones for nonlinear oscillators, described in Appendix A.2, provides the theoretical foundation for the existence of such mappings (A. T. Winfree 1974; Kuritz et al. 2018a). By observing changing dynamics over time one may furthermore trace the rewiring of cellular regulatory networks in response to treatments at different cellular contexts, e.g. cell cycle stages.

Part II

Control

Chapter 5

Passivity-based ensemble control for cell cycle synchronization

This chapter is based on the publications:

Karsten Kuritz et al. (2018a). 'Passivity-based ensemble control for cell cycle synchronization'. In: *Emerg. Appl. Control Syst. Theory*. Ed. by Roberto Tempo et al. 1st ed. Springer International Publishing. DOI: 10.1007/978-3-319-67068-3_1.

Karsten Kuritz et al. (2018b). 'Ensemble control for cell cycle synchronization of heterogeneous cell populations'. In: *IFAC-PapersOnLine* 51.19. 7th Conference on Foundation of Systems Biology in Engineering FOSBE 2018, pp. 44–47. DOI: 10.1016/j.ifacol.2018.09.034.

In this Chapter, we investigate the problem of synchronizing a population of cellular oscillators in their cell cycle. Synchronizing a growing population in its cell cycle provides new opportunities for cell culture in basic research and applied biotechnology, in particular, if the populations is synchronized while hardly affecting individual cells in their natural cell cycle. Restrictions on the observability and controllability of the population imposed by the nature of cell biology give rise to an ensemble control problem specified by finding a broadcast input based on the distribution of the population. We solve the problem by a passivity based control law which we derive from the reduced phase model representation of the population and the aim of sending the norm of the first circular moment to one. Furthermore, we present conditions on the phase response curve and circular moments of the population which are necessary and sufficient for synchronizing a population of cellular oscillators.

This chapter is taken in parts from Kuritz et al. (2018a) and Kuritz et al. (2018b).

5.1 Background and problem formulation

The cell cycle is central to life. Every living organism relies on the cell division cycle for reproduction, tissue growth and renewal. Malfunction in this highly controlled cell cycle machinery is linked to various diseases, including Alzheimer's disease and cancer (Hanahan and Weinberg 2011; Zhivotovsky and Orrenius 2010). Cause and cure of these diseases are two sides of the same coin, and thus understanding of the cell cycle machinery and approaches to control it are subjects of ongoing research (Morgan 2007). Mathematically, the cell cycle machinery can be described as a dynamical system which obeys limit cycle behaviour (Csikasz-Nagy 2009; Claude Gérard et al. 2012) with dynamics of the general form

$$\dot{x} = f(x, u) \,. \tag{5.1}$$

Therein, the states x represent different molecular species in the cell which can be indirectly affected by external inputs u such as growth conditions, drugs and other environmental factors. Another control approach can be realized by directly regulating the expression levels of specific proteins, e.g. by optogenetics (Levskaya et al. 2010). Besides the agent based description, with each agent being a cellular oscillator with dynamics by Eq. (5.1), proliferating cell populations are often represented by structured population models (Jean Clairambault et al. 2007; Gyllenberg and Webb 1990). The resulting dynamics are governed by partial differential equations, belonging to the class of *Liouville equations* (Brockett 2012) of the general form

$$\partial_t p(x,t) = -\langle \partial_x, f(x,u) p(x,t) \rangle \,. \tag{5.2}$$

The concept of reduced phase models, reviewed in Appendix A.2, connects the nonlinear dynamics in Eq. (5.1) with age-structured population models, thereby facilitating control approaches based on the phase distribution of nonlinear oscillators (Kuramoto 1984; Mirollo and Strogatz 1990). Control of these oscillators is studied intensively, e.g. by the authors of (Wilson and Jeff Moehlis 2014; Montenbruck et al. 2015; Scardovi et al. 2010).

Several constraints imposed by the nature of cell biology complicate the task to synchronize a cell population in its cell cycle: (1) Experimental observation of the cell cycle state of individual agents over time is barely possible. A more realistic experimental observation is composed of representative samples drawn from the population from which the distribution of cells in the cell cycle must be reconstructed (Kuritz et al. 2017; Zeng et al. 2016). (2) Two new agents arise by division at the end of the cell cycle, resulting in exponential growth of the number of controlled agents and non smooth boundary conditions of the PDE. (3) Only broadcast input signals can be realized, giving rise to the following ensemble control problem:

Problem 5.1. Given a representative samples of the population as measurement, find a broadcast input signal u for a population of cellular oscillators such that the agents synchronize in their cell cycle.

Our approach to solve the above stated control problem is organized as follows. Section 5.2 briefly introduces the theoretic foundation of our control approach, compromising the classical input-/output framework for passivity based controller design and reduced phase models for the representation of weakly coupled oscillators. The control methodology is developed in Section 5.3. Section 5.4 examines the control methodology applied to a nonlinear ODE model of the mammalian cell cycle. Section 5.5 contains concluding comments.

5.2 Theoretical foundation

As mentioned above, we are interested in controlling a population of many identical uncoupled dynamical systems Eq. (5.1). The dynamics of the population follows the aforementioned Liouville equation Eq. (5.2), so for a given input $u(t)$ and initial distribution $p(\xi, 0) = p_0(\xi)$ we may find the solution of the PDE

$$p(\cdot, t) = \Upsilon(u, p_0, t) \,, \quad t > 0 \,. \tag{5.3}$$

An observable feature may for instance be the moments of Eq. (5.3), and the output of the system may be any function of these moments. More general, we consider any function which maps the solution of the PDE to a scalar value as a possible output function

$$y(t) = h(p(\cdot, t)), \quad y(t) \in \mathbb{R}. \tag{5.4}$$

By doing so, we establish an input-/output relation

$$y = H u \tag{5.5}$$

where H is some mapping that specifies y in terms of u. We will develop our control methodology based on the fundamentals of classical input-/output frameworks and the concept of reduced phase models, reviewed in Appendix A.

5.2.1 From reduced phase model to age-structured population models

Nonlinear oscillating systems are often studied by transforming the complex dynamic equations that describe their behavior into a phase coordinate representation (Kuramoto 1975; Arthur T Winfree 1967). This approach, reviewed in Appendix A, yields simplified yet accurate reduced phase models that capture essential properties of an oscillating system with a stable periodic orbit and is especially compelling from a control-theoretic perspective (Kuramoto 1984; Mirollo and Strogatz 1990). Given a family of oscillators in its reduced phase representation Eq. (A.14), the corresponding Liouville equation for the time evolution of the number density $n(x,t)$ of oscillators on the unit circle reads

$$\partial_t n(x,t) + \partial_x (\kappa(x,u) n(x,t)) = 0. \tag{5.6}$$

The control vector field equals the reduced phase model $\kappa(x,u) = \omega + Z(x)u$. In case of a cell population, a division of a mother cell into two daughter cells results in the boundary condition

$$n(0,t) = 2\, n(2\pi, t). \tag{5.7}$$

The model Eq. (5.6) and Eq. (5.7), with $u(t) = 0$, belongs to the model class of age-structured population models, based on the well known von Foerster-McKendrick models (McKendrick 1926; Foerster 1959), which are widely used to study cell cycle-related processes. In these models, the distribution of cells $q(x,t) = n(x,t)/N(t)$, obtained by normalizing the number density with the total cell number $N(t) = \int_0^{2\pi} n(x,t)\, \mathrm{d}x$ admits a time invariant distribution

$$\bar{q}(x) = 2\gamma\, \mathrm{e}^{-\gamma x}, \tag{5.8}$$

where $\gamma = \omega \log 2 = \frac{\log 2}{T}$ is the growth rate of the population (Powell 1956).

We will characterize a circular distribution by its Fourier coefficient or similarly its circular moments, as reviewed in Appendix A.3. In a synchronized population corresponding to a Dirac delta distribution, the L_1-norm of the first circular moment $|m_1| = r$ is equal to one. The control problem to synchronize (or balance) the agents in the population can now be stated as:

Problem 5.2. Given the system defined by Eqs. (5.6) and (5.7), find a control input u, such that $|m_1(q(\cdot,t))| \to 1 (\text{or } 0)$.

5.3 Ensemble control for cell cycle synchronization

This section states the control algorithm which we formulate in an input-/output framework, reviewed in Section A.1.2. We will first elaborate how to choose an output function h such that it is connected to our goal of synchronizing (or balancing) the population of agents. At the same time the mapping $H_1 : u \mapsto y$ is passive under this choice of output and by applying a strictly passive controller H_2 in the control approach of Figure 5.1, we conclude that the L^2-norm of y is finite ($y \in \mathcal{L}$). We will then study invariance properties and conditions of our system under the proposed control law. An interpretation of this result along with some further considerations indicate that the control law indeed synchronizes (or balances) the population of agents, thereby solving Problem 5.2.

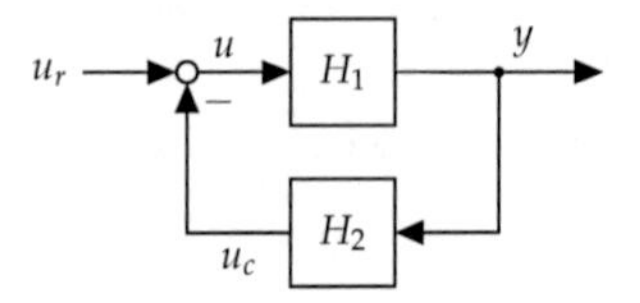

Figure 5.1. Control approach in the input-/output framework. If the mapping H_1 is passive and the controller H_2 is strictly passive one concludes that $y \in \mathcal{L}$.

5.3.1 Enabling passivity based controller design

The controller design based on the theory of passive systems benefits from a system model for which the control objective remains constant whenever $u = 0$. This property is not met by the model Eq. (5.6). In the following section we propose state transformations $n(x,t) \to p(x,t)$ such that $|m_1(p(\cdot,t))|$ remains constant whenever $u = 0$. The first transformation where we divide the number density in Eq. (5.6) by the steady-state distribution Eq. (5.8) eliminates the discontinuity at the boundary. We thus define $\tilde{n}(x,t) = n(x,t)/\bar{q}(x)$, resulting in

$$\begin{aligned} \partial_t \tilde{n}(x,t) + \partial_x \left(\kappa(x,u)\tilde{n}(x,t)\right) &= \gamma\kappa(x,u)\tilde{n}(x,t) , \\ \tilde{n}(0,t) &= \tilde{n}(2\pi,t) . \end{aligned} \tag{5.9}$$

Next, we define $p(x,t) = \tilde{n}(x,t)/\int_0^{2\pi} \tilde{n}(x,t)\,\mathrm{d}x$ such that $\int_0^{2\pi} p(x,t)\,\mathrm{d}x \equiv 1$, resulting in a proper probability distribution with PDE

$$\begin{aligned} \partial_t p(x,t) + \partial_x \left(\kappa(x,u)p(x,t)\right) &= u\gamma p(x,t)\left(Z(x) - \int_0^{2\pi} Z(x)p(x,t)\,\mathrm{d}x\right) , \\ p(0,t) &= p(2\pi,t) . \end{aligned} \tag{5.10}$$

The system Eq. (5.10) has now the favorable properties that facilitate the feedback approach for synchronization of the population: (1) $p(x,t)$ is a probability distribution with $\int_0^{2\pi} p(\xi,t)\,\mathrm{d}\xi \equiv 1$, (2) $p(x,t)$ is smooth over the boundary and (3) the length of the first circular moment $|m_1(p(\cdot,t))|$ remains constant whenever $u = 0$ as this results in the PDE for a travelling wave

$$\begin{aligned} \partial_t p(x,t) + \omega\,\partial_x p(x,t) &= 0 \\ p(0,t) &= p(2\pi,t) . \end{aligned} \tag{5.11}$$

Furthermore, if $|m_1| = 1$, then the agents are synchronized.

5.3.2 Synchronization of the population

With the model Eq. (5.10) given, it remains to define an appropriate output and a suitable output feedback control law which synchronizes the population. This will be achieved by choosing the output $y = h(p(\cdot, t))$ such that: (1) $y = 0$ whenever the population is synchronized, and (2) the map $H_1 : u \mapsto y$ is passive. To increase clarity in the derivation, we use a shorthand notation where we omit the argument in the functional $m_1((p(\cdot, t))$. As synchrony is equivalent to $|m_1| = 1$, we first study the time derivative of $|m_1(p(\cdot, t))|$ evolving under Eq. (5.10), viz.

$$\begin{aligned}\frac{\mathrm{d}}{\mathrm{d}t}|m_1(p(\cdot,t))| = \Big((\gamma+\mathrm{i})m_{-1}\int_0^{2\pi} \mathrm{e}^{\mathrm{i}x} Z(x)p(x,t)\,\mathrm{d}x \\ - 2\gamma m_1 m_{-1}\int_0^{2\pi} Z(x)p(x,t)\,\mathrm{d}x + (\gamma-\mathrm{i})m_1\int_0^{2\pi} \mathrm{e}^{-\mathrm{i}x} Z(x)p(x,t)\,\mathrm{d}x\Big)\, u\,.\end{aligned} \tag{5.12}$$

In the following, we define $d^p(x) = d(p(\cdot, t), x)$ with

$$\begin{aligned}d(p(\cdot,t),x) &= \sum_{l=-1}^{1} d_l\, \mathrm{e}^{\mathrm{i}lx}\,, \\ d_{-1} = (\gamma-\mathrm{i})m_1\,, \quad d_0 &= -2\gamma m_1 m_{-1}\,, \quad d_1 = (\gamma+\mathrm{i})m_{-1}\,.\end{aligned} \tag{5.13}$$

This leads to a more practical representation of Eq. (5.12) in terms of the inner product

$$\frac{\mathrm{d}}{\mathrm{d}t}|m_1(p(\cdot,t))| = \langle Z, d^p p(\cdot,t)\rangle\, u\,, \tag{5.14}$$

which is zero whenever $u = 0$. Thus, by choosing the output as

$$h(p(\cdot,t)) = \langle Z, d^p p(\cdot,t)\rangle\,, \tag{5.15}$$

we arrive at the following observations.

Lemma 5.1. *The mapping H_1 given by Eq. (5.5) with output Eq. (5.15) and internal dynamics Eq. (5.10) is passive.*

Proof. Following the definition of Desoer and Vidyasagar (1975), the system is passive if $\langle y, u\rangle_T \geq \beta$, $\forall u \in \mathcal{L}_e$, $\forall T \in \mathbb{R}^+$. We constructed y such that

$$\langle y,u\rangle_T = \int_0^T y(t)u(t)\,\mathrm{d}t = \int_0^T \frac{\mathrm{d}}{\mathrm{d}t}|m_1(p(\cdot,t))|\,\mathrm{d}t = |m_1(p(\cdot,T))| - |m_1(p(\cdot,0))|$$

and as the norm of the first circular moment of a probability distribution is upper bounded by 1, we can choose $\beta = -1$. ■

Theorem 5.1. *If the output $y(t)$ is given by Eq. (5.15), and the output feedback is chosen as $u(t) = -y(t)$, then the output $y(t)$ converges to zero.*

Proof. The result follows directly from the basic passivity theorem given in Desoer and Vidyasagar (1975), namely that the output of a passive system H_1 lies in $\mathcal{L}$ if the output is fed back through a strictly passive system H_2. By Lemma 5.1, H_1 is a passive system. Further, we chose H_2 as the identity function $H_2 x = x$, which indeed is strictly passive, and $y \in \mathcal{L}$. With y being uniform continuous we know from Barbalat's Lemma, that $y(t) \to 0$, thereby concluding the proof. ■

Next, we study the invariance properties of our system having zero output. This is necessary as we want to exclude the existance of invariant distributions other than synchrony. Our study is based on the properties of Fourier series and the Fourier coefficients of Z, d^p and $p(\cdot,t)$ in Eq. (5.15). The Fourier series of a function $F : x \mapsto F(x)$ and its Fourier coefficients in any 2π-interval are given by

$$F(x) = \sum_{k=-\infty}^{\infty} a_k \, \mathrm{e}^{\mathrm{i}kx}, \quad \text{with } a_k = \frac{1}{2\pi} \int_0^{2\pi} \mathrm{e}^{-\mathrm{i}kx} F(x) \, \mathrm{d}x \,. \tag{5.16}$$

To keep the notation in accordance with the definition of the circular moments in Eq. (A.15), we introduce a modified series representation and scaled coefficients as

$$F(x) = \sum_{k=-\infty}^{\infty} \frac{b_k}{2\pi} \mathrm{e}^{-\mathrm{i}kx} \quad \text{with } b_k = \int_0^{2\pi} \mathrm{e}^{\mathrm{i}kx} F(x) \, \mathrm{d}x \,. \tag{5.17}$$

A distribution $p(\cdot,t)$ is contained in a forward invariant set in $\mathcal{E} = \{p\colon h(p) = 0\}$ if and only if $h(p(\cdot,t+\tau)) = 0$, $\forall \tau \geq 0$. The modified series representation leads to the following lemma:

Lemma 5.2. *Let c_k, d_k and m_k be the coefficients of Z, d^p and $p(\cdot,t)$, respectively. Then $\mathcal{E}$ is invariant if and only if*

$$k c_k \nu_k = 0 \,, \quad \forall k \in \mathbb{Z} \,, \tag{5.18}$$

with $\nu_k = \sum_{l=-1}^{1} d_l m_{k-l}$.

Proof. Due to the periodicity of the cell cycle we know that E is in an invariant set if and only if $h(p(\cdot,t+\tau)) = 0$, $\forall \tau \in [0,T]$, which is due to the constant propagation ($u = 0$) of $p(x,t)$ with $\frac{\mathrm{d}x}{\mathrm{d}t} = \omega$ equal to $h(p_{\omega\tau}(\cdot,t)) = h(p_\sigma(\cdot,t)) = 0$, $\forall \sigma \in [0,2\pi]$, where we define $p_\sigma(x,t) = p(x-\sigma,t)$. We will use this notation to denote a shift in x also for Z later on. If $h(p_\sigma(\cdot,t)) = 0$, then this is also true for its derivative $\frac{\mathrm{d}}{\mathrm{d}t} h(p_\sigma(\cdot,t))$ resulting in the following condition for invariance

$$\frac{\mathrm{d}}{\mathrm{d}t} \langle Z, d^{p_\sigma} \, p_\sigma(\cdot,t) \rangle = 0 \,, \quad \forall \sigma \in [0,2\pi] \,. \tag{5.19}$$

The derivative is obtained by employing the identity from the PDE Eq. (5.10) with $u = 0$: $\partial_t p(x,t) = -\omega \, \partial_x p(x,t)$ and subsequently integrating by parts. Furthermore, the shift in x is transferred to the PRC by a change of variables $x = \xi + \sigma$, changing Eq. (5.19) to

$$\langle \frac{\mathrm{d}}{\mathrm{d}x} Z_{-\sigma}, d^p \, p(\cdot,t) \rangle = 0 \,, \quad \forall \sigma \in [0,2\pi] \,. \tag{5.20}$$

The last steps of the proof are: (1) substituting $p(\cdot,t)$ and $Z_{-\sigma}$ with its modified Fourier series and (2) employing Parseval's theorem. With c_k being the coefficients of Z, the coefficients of the argument shifted derivative $\mathrm{d}\, Z_{-\sigma} / \, \mathrm{d}\, x$ in Eq. (5.20) are $-\mathrm{i}\, k \, \mathrm{e}^{-\mathrm{i}k\sigma} c_k$. The function d^p has Fourier coefficients d_{-1}, d_0, d_1 and all other coefficients equal zero. The product $d^p p(\cdot,t)$ has modified coefficients ν_k. By Parseval's theorem, the inner product in Eq. (5.20) equals the sum of its coefficients

$$\langle \frac{\mathrm{d}}{\mathrm{d}x} Z_{-\sigma}, d^p \, p(\cdot,t) \rangle = \frac{-i}{(2\pi)^2} \sum_{k=-\infty}^{\infty} \mathrm{e}^{-\mathrm{i}k\sigma} \, k c_k \nu_k \tag{5.21}$$

which can be written as inner product, and therefore the condition for invariance is

$$\langle (\mathrm{e}^{-\mathrm{i}k\sigma})_k, (kc_k v_k)_k \rangle = 0\,, \quad \sigma \in [0, 2\pi]\,, k \in \mathbb{Z}\,. \tag{5.22}$$

The series $\left(\mathrm{e}^{-\mathrm{i}k\sigma}\right)_k$ are basis functions of a complete orthogonal basis, hence the inner product Eq. (5.22) is zero if and only if $kc_k v_k = 0$, $\forall k \in \mathbb{Z}$. This equals Eq. (5.18), thereby concluding the proof of Lemma 5.2. ■

Lemma 5.2 relates the condition for invariance where $y(t) \equiv 0$ to the Fourier coefficients of the phase response curve and the circular moments of the distribution of cell. With Lemma 5.2 at hand we can thus identify conditions on the phase response curve Z, such that the synchronized and balanced population are the only invariant ones.

Theorem 5.2. *If the output feedback $u(t) = -y(t)$ is chosen for system H_1 and the output $y(t)$ is given by Eq. (5.15), then*

$$\begin{aligned} \mathcal{M}_0 &= \{p\colon |m_1(p)| = 0\}\,, \\ \mathcal{M}_1 &= \{p\colon |m_1(p)| = 1\} \end{aligned}$$

are invariant sets in $\mathcal{E}$. Furthermore, if the first moment of Z is not equal to zero, i.e. $c_1 \neq 0$, then no other invariant set exists.

Proof. By Lemma 5.2, invariance of $\mathcal{E}$ requires Equation Eq. (5.18) to be fulfilled. Invariance of $\mathcal{M}_0$ and $\mathcal{M}_1$ is then verified by showing that $v_k = 0$, $\forall k \in \mathbb{Z}$. As $|m_1| = 0$ implies $m_1 = m_{-1} = 0$, Eq. (5.18) is trivially met, and $\mathcal{M}_0$ is invariant. If $|m_1| = 1$, then all moments are given by $m_k = \mathrm{e}^{ik\phi}$ and thus have their L_1-norm equal to one. All terms in v_k cancel out, hence $\mathcal{M}_1$ is invariant. To conclude the proof of Theorem 6.1 we verify that $c_1 = 0$ is a necessary condition for Eq. (5.18) by showing that $v_1 \neq 0$ whenever $|m_1| \notin \{0, 1\}$. m_1 and m_{-1} are again represented as complex numbers. Furthermore $p(\cdot, t)$ is a probability distribution with $m_0 = 1$ by definition and we get

$$v_1 = r\left(\mathrm{e}^{-\mathrm{i}\phi}(\gamma + \mathrm{i}) + \mathrm{e}^{\mathrm{i}\phi}\left((\gamma - \mathrm{i})m_2 - 2\gamma r\right)\right)\,. \tag{5.23}$$

From $|m_1| = r \neq 0$, one realizes that $\mathrm{e}^{-\mathrm{i}\phi}$ and $\mathrm{e}^{\mathrm{i}\phi}$ are orthogonal and $\gamma > 0$. By definition, it follows that $v_1 \neq 0$. Hence, $\mathcal{M}_0$ and $\mathcal{M}_1$ are the only invariant sets in $\mathcal{E}$ if $c_1 \neq 0$. ■

We will now discuss some aspects regarding the convergence to a synchronized (or balanced) population, given that the first moment of the phase response is not equal to zero. If the output is given by Eq. (5.15) and the output feedback $u(t) = -\varepsilon\, y(t)$, $\varepsilon > 0$, is chosen for system H_1, then

$$\frac{\mathrm{d}}{\mathrm{d}t}|m_1(p(\cdot, t))| = -\varepsilon\, h(p(\cdot, t))^2 \leq 0\,, \qquad \forall t \geq 0\,, \tag{5.24}$$

and $|m_1(p(\cdot, t))|$ decreases monotonically. Furthermore, the average of Eq. (5.24) over one period is strictly monotonically decreasing whenever $|m_1| \notin \{0, 1\}$. These observations suggest that $|m_1(p(\cdot, t))|$ approaches $\mathcal{M}_0$ from almost all initial conditions and the population is balanced. Synchronization of the population is achieved by sign reversal of the output function $y(t) = -h(p(\cdot, t))$ and the same output feedback. Sign reversal of the output preserves passivity of the system and by $\frac{\mathrm{d}}{\mathrm{d}t}|m_1(p(\cdot, t))| \geq 0$, p approaches $\mathcal{M}_1$ and a synchronized distribution is achieved.

Remark 5.1. Theorems 5.1 and 6.1 and the fact that an attractive set becomes a repelling set by sign reversal, strongly suggest that the population with output Eq. (5.15) and control input $u(t) = \varepsilon y(t)$ converges to a Dirac delta distribution. However, due to topological reasons, analysis of convergence of $p(\cdot, t)$ is difficult and beyond the scope of the present study.

This section derived the ensemble control algorithm to synchronize a growing cell population in its cell cycle. The population-level feedback defined by $u(t) = -y(t)$ with output $y(t)$ given by Eq. (5.15) requires knowledge of the phase response curve and measurements of the distribution of cells over the cell cycle. The cell cycle distribution can be calculated from single-cell snapshot data by transforming the pseudotemporal ordering to an age scale as described in Chapters 2 and 3. We present such an approach in the simulation studies in the next section.

5.4 Computational studies

We will next examine the developed control methodology extensively in two simulation studies. First, we demonstrate the controller for the reduced phase model Eq. (5.10) where we show convergence of the population to a synchronized or balanced population. Second, we examine controller performance with a multi-agent setup. Therein, simulating an exponentially growing cell population in an individual-based setting challenges the controller in a highly realistic scenario.

We used a five-variable skeleton model of the mammalian cell cycle by C. Gérard and Goldbeter (2011) and Claude Gérard et al. (2012) consisting of the four main cyclin/Cdk complexes, the transcription factor E2F and the protein Cdc20. A detailed description of the model together with the model equations is given in Appendix D.

5.4.1 Reduced phase model simulation

We now demonstrate the developed control methodology on the reduced phase model Eq. (5.10). We extended the model by an additive input to the dynamics of Cyclin A

$$\dot{\xi}_{\mathrm{CycA}} = f_{CycA}(\xi) + \frac{1.6\,(\alpha - \xi_{CycA})}{0.1 + \alpha - \xi_{CycA}}\, u(t)\,. \tag{5.25}$$

The input can be thought of e.g. an optogenetic signal causing a direct induction of Cyclin A expression with the total amount of Cyclin A being upper bounded by α. The phase response curve Z was obtained by solving the appropriate adjoint equation (see Appendix A.2) using the dynamic modeling program XPPAUT (Malkin 1956; Ermentrout 2002). We take u according to Theorem 5.1 and simulated both the synchronizing and balancing scenario with h as defined in Eq. (5.15). The results are depicted in Figure 5.2. In the synchronizing scenario, one observes how the first moment approaches the unit circle, indicating that the distribution of cells indeed converges to the Dirac distribution. This can also be observed in the simulation snapshots. By sign reversal of the output and starting with a imbalanced cell density, we further see that this process is reversed and the population approaches a uniform distribution.

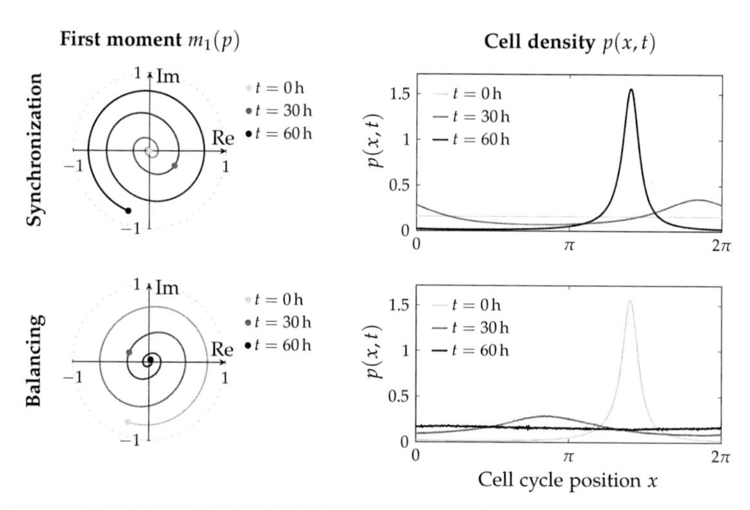

Figure 5.2. Simulation of Eq. (5.10) derived from a 5-state cell cycle model with both a synchronizing (top) and a balancing (bottom) controller. On the left: temporal evolution of the first circular moment m_1 in the complex plane. On the right: snapshots of the cell density over the cell cycle position

5.4.2 Realistic scenario in an individual-based simulation framework

The derived output feedback and its stability analysis rely on several assumptions which may restrict the applicability of the approach. Most important, the controller is based on a PDE of the reduced phase model representation of the population (Figure A.2). In this idealized description, the complex dynamics of oscillators on a limit cycle are captured by a single dynamic on the unit circle. The theory is based on the concept of weakly coupled systems (Hoppensteadt and Eugene M. Izhikevich 1997) which imposes certain constraints on the input strength.

In this section we investigate the performance of the control law in an *in silico* study where we simulate a heterogeneous cell population with an individual-based simulation framework (Imig et al. 2015). Therein, the dynamics of each cell are captured by ordinary differential equations. Resampling of cellular states at cell division reflects the naturally occurring heterogeneity in a population due to uneven distribution of proteins during cytokinesis (Huh and Paulsson 2011). Performance of the controller is examined in different scenarios where we vary variance during resampling against the input strength. Our results indicate a suitable range for input strength where synchrony is achieved despite the naturally occurring heterogeneity.

Implementation

We will first describe the dynamical model and the necessary steps to calculate the input from the full state model by C. Gérard and Goldbeter (2011). We then outline the implementation of the controller in an individual-based simulation framework for heterogeneous populations (Imig et al. 2015).

The underlying ODE model is a 5-state skeleton model of a mammalian cell cycle (Figure D.1 a, Appendix D). For the purpose of controlling the cell cycle progression, we again extended the dynamics of the cyclin A/Cdk2 complex by an additive input

$$\dot{\xi}_{\mathrm{CycA}} = f_{CycA}(\xi) + 1.6\left(1 - \frac{\xi_{CycA}^8}{4^8 + \xi_{CycA}^8}\right)\epsilon\, u\,, \tag{5.26}$$

causing a direct induction of cyclin A expression. At high levels of cyclin A the effect of the optogenetic input rapidly declines mimicking saturation of the amount of cyclin A. The input strength is tuned by the factor ϵ which is essential in order to keep the oscillators in a neighborhood W of the limit cycle γ. For the calculation of the input $u(t)$ the phase of each agent has to be determined. This process follows the theory of reduced phase models, where $\psi_2 : W \to \gamma$ maps a point $\xi(t) \in W$ to the generator of its isochron $\xi(t) \in \gamma$ (Figure A.2). However, due to the high complexity of the notion of isochrons we used a simplified mapping where $\hat{\psi}_2$ maps the point $\xi(t)$ to its closest point on the limit cycle $\hat{\xi}(t) \in \gamma$. The relation $\psi_1 : \gamma \to \mathcal{S}^1$ which maps a point on the limit cycle γ to its corresponding position on the unit circle $\mathcal{S}^1$ which is proportional to cell age can easily be obtained by inversion of the simulated trajectory on the limit cycle $\xi^{-1}(\gamma) = x/\omega$. The composition $\psi = \psi_1 \circ \hat{\psi}_2$ of the two mappings now maps the point $\xi(t) \in W$ uniquely to its phase $x(t) \in \mathcal{S}^1$ on the unit circle. In an *in vitro* experiment this mapping can be achieved by employing ergodic principles as described in Chapters 2 and 3 or in Kuritz et al. (2017) and Kuritz et al. (2020b). The PRC Z was obtained by solving the appropriate adjoint equation (see Appendix A.2) using the dynamic modeling program XPPAUT (Malkin 1956; Ermentrout 2002).

The distribution of cells on the unit circle was calculated by kernel density estimation from phase values x_k using a von Mises distribution as kernel with manually tuned bandwidth. The input is calculated from the PRC and the distribution of cells on the unit circle as stated in Theorem 5.1. The simulation framework is based on the individual-based population model by Imig et al. (2015) and was slightly adapted, resulting in the following framework (Figure 5.3):

- The simulation runs from time-points t_0 until t_{end}.
- A list *cells* contains all initial mother-cells at t_0.
- The cell with the shortest simulated time t_0 is simulated until a specified time $\hat{t}_{end}$, or until the cell complies with a division condition.
- Upon cell division, heterogeneity is introduced by sampling the initial condition of the two daughter cells from a logarithmic normal distribution with sampling variance σ and the mean equal to the state values at the limit cycle γ at cell division.
- The total number of cells is kept constant by removing a randomly determined cell from the list *cells* whenever a cell divides.

A snapshot measurement is taken at the end of each simulation loop with a simulation time of $\hat{t}_{end} = 1\,h$ from which the input for the next loop is calculated.

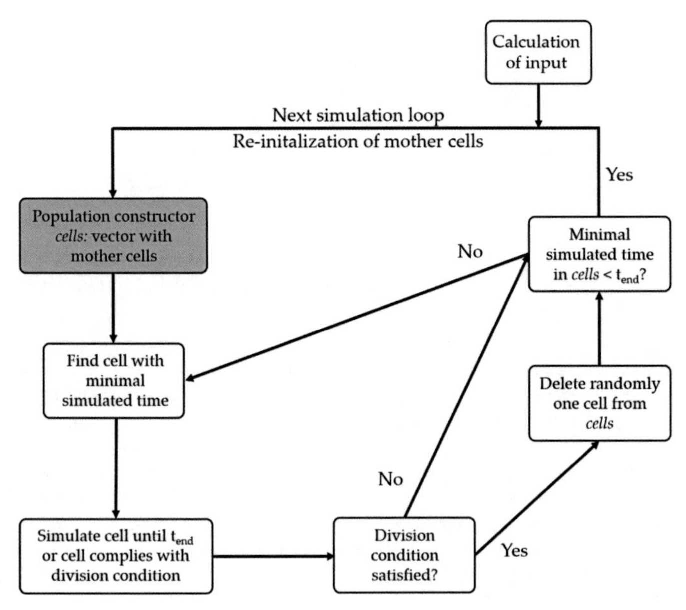

Figure 5.3. Scheme of the first layer of the simulation framework (adapted from Imig et al. (2015)).

Simulation results

The cell population was initialized by applying the inverse mapping $\psi_1^{-1} : \mathcal{S}^1 \rightarrow \gamma$ on samples drawn from a von Mises distribution with circular moment $m_1(t_0) = 0.3$. Simulations with varying input factor ϵ and sampling variance σ were run for 500 hours. Exemplary results obtained for sampling variance of 0.13 and 0.18 and input strength varying between 0 and 0.09 illustrate the different effects of the controller on the cell population (Figure 5.4).

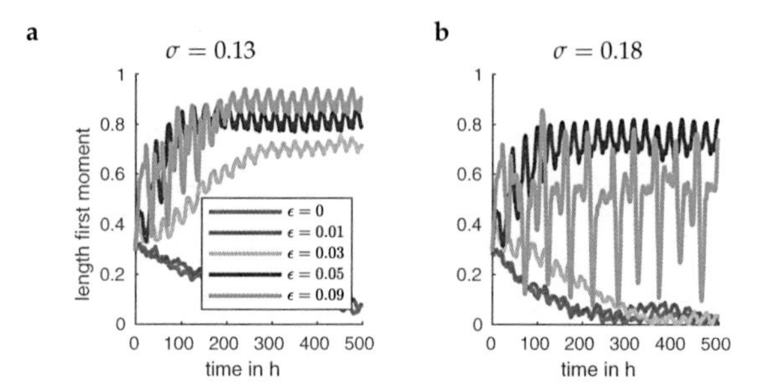

Figure 5.4. Simulation results showing the absolute value of the first circular moment for sampling variances $\sigma = 0.13$ (**a**) and $\sigma = 0.18$ (**b**) and input strength varying between 0 and 0.09.

For $\epsilon < 0.02$, the input is too small and the norm of the first circular moment remains constant or decreases, indicating a conversion towards a steady-state distribution. Increasing the input strength to $\epsilon = 0.05$ results in stable oscillations with increasing length of the first moment. However, input strengths $\epsilon \geq 0.09$ are too strong causing the cells to diverge from their original limit cycle which results in loss of oscillation of parts of the population. Cell cycle arrest of a fraction of the population results in a very large amplitude of the length of the first moment. When the oscillating subpopulation passes by the arrested cells $|m_1|$ increases. When the cells are on opposite sides of the cell cycle, $|m_1|$ is small. Such a behaviour can be seen for $\epsilon = 0.09$ at the early time points with $\sigma = 0.13$ and throughout the whole simulation with $\sigma = 0.18$ (Figure 5.4).

As illustrated in Figure 5.5, input strength and sampling variance were varied over a sufficiently large range to cover relevant outcomes and define regions where: (1) the input predominates over noise and the population synchronizes, (2) the population diverges and noise introduced during cell division predominates, or (3) the input is too strong causing loss of oscillations. The input strength which is necessary to synchronize the population increases with the variance until it reaches a point where the input causes loss of oscillations. The border of this region expands at higher variances to lower input strengths. A possible explanation for this observation is that higher variances during resampling result in larger deviations from the limit cycle during cell division such that lower inputs suffice to interrupt stable oscillations. The decline rate of $|m_1|$ in the simulation with zero input and sampling variances of $\sigma = 0.2$ is comparable to a cell cycle length distribution with mean 24 h and variance 2 h. Such a distribution is in good accordance with the experimentally observed cell cycle lengths of mammalian cell cultures.

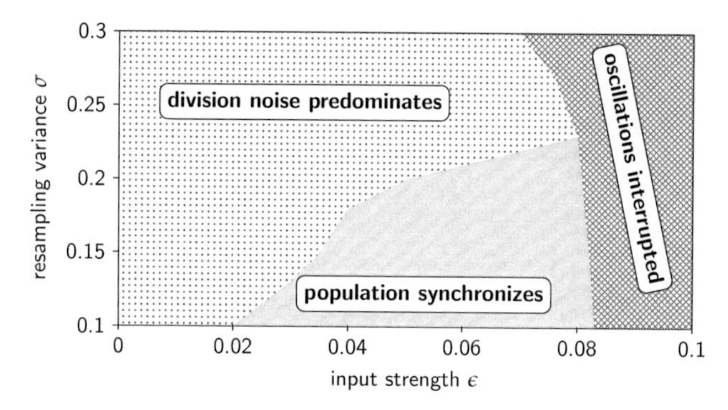

Figure 5.5. Parameter regions where we observe no synchronization due to predominant division noise, successful synchronization or loss of oscillations. Dominance of division noise and successful synchronization are characterized by the first moment apporaching values smaller or larger than 0.5, respectively. Loss of oscillations is characterized by large amplitudes of the first moment over the course of the simulation.

5.5 Summary and discussion

Summary

We studied the ensemble control problem of synchronizing a cell population in their cell cycle with restriction of the observation to representative samples of the population. Starting with a single cell as oscillator on a limit cycle, we developed a reduced phase model of the population with a broadcast input acting via the phase response curve. We then proposed state transformations for the age-structured population type model which enable controller design in the input-/output framework for passive systems. Formulating the control problem in terms of the first circular moment of the population led to the desired output feedback which synchronizes the population. Finally, we derived necessary and sufficient conditions on the phase response curve for the synchronization of the population. We expanded our results derived for the idealized behavior of the cell population to more realistic conditions in a simulation study. To this end we used an individual-based simulation framework where heterogeneity is introduced by resampling of states upon cell division (Imig et al. 2015). The simulation results indicate the existence of a suitable range in which the input strength can be varied while still achieving synchrony without interrupting the oscillations.

Discussion and Outlook

Heterogeneity in the population can also be introduced by stochastic dynamics. We hypothesize that the proposed control strategy is robust such that synchronization might also be achieved under stochastic dynamics. Typically, synchronization of a cell population is achieved by environmental or chemical treatments. This procedure is often accompanied by complex side effects and the population loses synchrony rapidly. The presented controller can be used to culture cell populations which are synchronized while hardly affecting individual cells in their natural cell cycle.

The present study solves the ensemble control problem of cell cycle synchronization by sending the first circular moment to one. However, we will show in the next chapter, that a similar approach is suitable to achieve any desired moment-determinate distribution by steering the circular moments of the population to the corresponding values of a target distribution.

Chapter 6

Ensemble control for cellular oscillators

This chapter is based on the publication:

> Karsten Kuritz et al. (2019). 'Ensemble Controllability of Cellular Oscillators'. In: *IEEE Control Systems Letters* 3.2, pp. 296–301. DOI: 10.1109/lcsys.2018.2870967. © 2019 IEEE

In this chapter, we investigate the control problem of achieving a desired distribution of cellular oscillators on the periodic orbit of their limit cycle. Many diseases including cancer, Parkinson's disease and heart diseases are caused by loss or malfunction of regulatory mechanisms of an oscillatory system. Successful treatment of these diseases might involve recovering the healthy behavior of the oscillators in the system, for example, achieving synchrony or a desired distribution of the oscillators on their periodic orbit. We consider the problem of controlling the distribution of a population of cellular oscillators described in terms of phase models. Different practical limitations on the observability and controllability of cellular states naturally lead to an ensemble control formulation in which a population-level feedback law for achieving a desired distribution is sought. A systems theoretic approach to this problem leads to Lyapunov- and LaSalle-like arguments, from which we develop our main contribution, novel necessary and sufficient conditions for the controllability of phase distributions in terms of the Fourier coefficients of the phase response curve. Since our treatment is based on a rather universal formulation of phase models, the results and methods proposed in this chapter are readily applicable to the control of a wide range of oscillating populations, such as circadian clocks and spiking neurons. Figure 6.1 illustrates the diverse objectives for our ensemble control algorithm in the case of cardiovascular diseases, sleep disorder and Parkinson's diseases.

This chapter is taken in parts from Kuritz et al. (2019),

6.1 Background and problem formulation

Periodic fluctuations in biological processes are found at all levels of life and are often the result of changes in gene expression. These rhythms play key roles in a variety of important processes, including the cell cycle, circadian regulation, metabolism, embryo development, neuron firing and cardiac rhythms (Arthur T. Winfree 1986; Hoppensteadt and Eugene M. Izhikevich 1997). Biological oscillators function as finely tuned dynamic systems in which

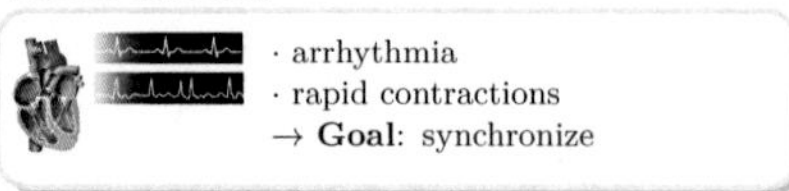

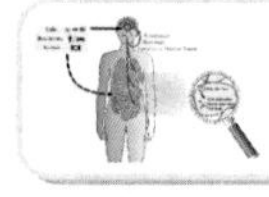

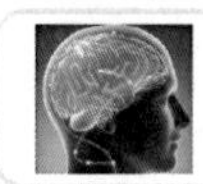

Figure 6.1. Cardiovascular diseases, sleep disorder and Parkinson's disease are related to malfunction in oscillatory systems. Successful treatment of these diseases might involve recovering the healthy behavior of the oscillators in the system, i.e., achieving synchrony, re-entrainment or desynchronization.

a time-delayed negative feedback gives rise to sustained rhythms. The rhythms are robust to noise while remaining acutely sensitive to various environmental and intracellular cues. Malfunction in these highly controlled oscillators is linked to various diseases, including Parkinson's and Alzheimer's disease, sleep disorder, cardiovascular diseases and cancer (Hanahan and Weinberg 2011; Zhivotovsky and Orrenius 2010; Mirsky et al. 2009). Cause and cure of these diseases are two sides of the same coin, and thus understanding oscillatory mechanisms and approaches to control it are subjects of ongoing research.

Mathematically, these oscillatory systems can be described as dynamical system with dynamics of the general form

$$\dot{x} = f(x, u) \,, \tag{6.1}$$

exhibiting a stable limit cycle (Hodgkin and Huxley 1952; Arthur T. Winfree 2001). Therein, the states x represent different molecular species in the cell which can be affected by external inputs u such as media, drugs, optogenetic approaches or environmental factors. Besides the agent-based description, with each agent being a cellular oscillator with dynamics Eq. (6.1), oscillating cell populations are modeled in terms of the distribution of cells in state space (Gyllenberg and Webb 1990). The resulting dynamics are governed by partial differential equations, belonging to the class of *Liouville equations* (Brockett 2012) of the general form

$$\partial_t p(t,x) = -\langle \partial_x, f(x,u)p(t,x) \rangle \,. \tag{6.2}$$

Nonlinear oscillating systems are often studied by transforming the complex dynamic equations that describe their behavior into a phase coordinate representation (Kuramoto 1975; Arthur T Winfree 1967). This approach, reviewed in Appendix A, yields simplified yet accurate reduced phase models that capture essential properties of an oscillating system with a stable periodic orbit and is especially compelling from a control-theoretic perspective (Kuramoto 1984; Mirollo and Strogatz 1990). Control design can be achieved for systems where the phase, but not the state, can be observed, and where the input response can be approximated experimentally when the dynamics are unknown (Wilson and Jeff Moehlis 2015; Wilson and Jeff Moehlis 2016; Kuritz et al. 2017).

Control of cellular oscillators is significant in biology, with a particular relevance in neuroscience (Eugene M Izhikevich 2007; Glass 2001). Applications in clinical medicine include protocols for coping with jet lag (Vosko et al. 2010; Carmona-Alcocer et al. 2018; Mirsky et al. 2009; J.-S. Li et al. 2013), clinical treatments for neurological disorders including epilepsy (Kiss et al. 2008) and Parkinson's disease (Hofmann et al. 2011; Wilson and Jeff Moehlis 2014; Matchen and J. Moehlis 2017; B. Monga et al. 2018). Furthermore, control of cellular oscillators takes place in cardiac pacemakers and during cancer treatment (Feillet et al. 2015). A common goal is to recover the healthy behavior of the oscillators, for example, by achieving synchrony or a desired distribution of the oscillators on their periodic orbit. The above mentioned applications are based on various theoretical results. For instance, optimal control approaches are widely studied with different focus based on the field of application (Matchen and J. Moehlis 2017; Zlotnik and J.-S. Li 2014). Furthermore, controllability of oscillators is studied within the framework of Lie algebra (Brockett 2010; Brockett 1976) where, for example, Lie brackets of drift and control vector fields are used to compute reachable sets for a population of non-identical oscillators (Zlotnik and J.-S. Li 2014). Recently, B. Monga et al. (2018) proposed a common control law for synchronizing and desynchronizing neural populations which is equivalent to the formulation from which we develop our controllability conditions. The framework is described in more general form in Bharat Monga and Jeff Moehlis (2019) together with an optimal control algorithm, to minimize control energy consumption.

As motivated in the introduction, we seek to find a broadcast input signal which steers the population of oscillators to a desired distribution on their limit cycle. Similarly as in Chapter 5, our control methodology will be based on the concept of reduced phase models, reviewed in Appendix A. We will now rigorously state the density control problem (see also B. Monga et al. (2018)) and relate it to a control problem for the moments of the population, cf. (Zeng and Allgöwer 2016; Kuritz et al. 2018a).

Given a family of weakly coupled identical oscillators in its reduced phase representation Eq. (A.14), the corresponding Liouville equation for the time evolution of the density, $p : \mathcal{S}^1 \times \mathbb{R}_{\geq 0} \to \mathbb{R}_{\geq 0}$, of oscillators on the unit circle reads

$$\partial_t p(x,t) + \partial_x \left(\kappa(x,u)p(x,t)\right) = 0 \tag{6.3}$$

with the boundary condition $p(0,t) = p(2\pi,t)$ and the initial condition $p(x,0) = p_0(x)$. Here, the (controlled) vector field is obtained from the reduced phase model, as

$$\kappa(x,u) = \omega + Z(x)u \,. \tag{6.4}$$

The target distribution, $q : \mathcal{S}^1 \times \mathbb{R}_{\geq 0} \to \mathbb{R}_{\geq 0}$, is supposed to be 2π-periodic with angular velocity ω and can thus be described by the PDE

$$\partial_t q(x,t) + \omega\, \partial_x q(x,t) = 0 \tag{6.5}$$

with the boundary condition $q(0,t) = q(2\pi,t)$ and the initial condition $q(x,0) = q_0(x)$. This PDE has the explicit solution $q(x,t) = q_0\left((x-\omega t) \mod 2\pi\right)$ which can be thought of as rotating initial density. We assume that $p_0(x)$ and $q_0(x)$ are (strictly) positive on $[0,2\pi]$, which results in the positivity of $p(x,t)$ and $q(x,t)$ for all $t \geq 0$. These definitions lead to the control problem as described similarly in B. Monga et al. (2018):

Problem 6.1. Given the system defined by Eq. (6.3) and Eq. (6.5), find a control input u depending on population-level data, such that the density $p(x,t)$ converges towards a desired periodic distribution $q(x,t)$.

6.2 Ensemble control for oscillator moments

This section derives the control methodology to solve Problem 6.1. The starting point will be the introduction of a natural cost functional V by which the control problem can be formulated as the minimization of this cost functional. In investigating the dynamics of V, we eventually arrive at the very favorable dynamic structure

$$\frac{\mathrm{d}}{\mathrm{d}t}V(t) = \phi(t)u(t)\,, \tag{6.6}$$

where ϕ incorporates population-level data. Thus, we are immediately led to the simple (population-level) feedback law $u(t) = -\phi(t)$, resulting in

$$\frac{\mathrm{d}}{\mathrm{d}t}V(t) = -\phi(t)^2 \leq 0\,. \tag{6.7}$$

A LaSalle-like argument then establishes the convergence $V(t) \to 0$ rigorously.

For better readability we will define a shorthand notation for frequently used expressions. Let $p(x,t)$ be a distribution function on the unit circle at time t. We will use $p_t(x) = p(x,t)$ especially when referring to the function in a L_2 inner product $\langle \cdot, \cdot \rangle$. For example, we denote the L_2-norm of $p(x,t)$ by

$$\langle p_t, p_t \rangle = \int_0^{2\pi} p(x,t)p(x,t)\,\mathrm{d}x\,.$$

6.2.1 Controller design for arbitrary distributions

Given the model Eq. (6.3) and Eq. (6.5), we introduce a cost functional that measures the distance between the actual density $p(x,t)$ and the reference density $q(x,t)$. It is natural to choose this cost functional as the L_2-norm of the difference between both distributions

$$V(t) = \frac{1}{2}\langle \Delta_t, \Delta_t \rangle \tag{6.8}$$

where $\Delta_t = p_t - q_t$. By considering the time derivative of this storage function we arrive at the following Lemma:

Lemma 6.1. *The time derivative of the storage function Eq.* (6.8) *under the dynamics given by Eqs.* (6.3) *and* (6.5) *is*

$$\frac{\mathrm{d}}{\mathrm{d}t}V(t) = \left(\int_0^{2\pi} (\partial_x \Delta_t) Z p_t \,\mathrm{d}x \right) u(t)\,. \tag{6.9}$$

Proof. The proof of Lemma 6.1 results from straightforward but somewhat involved computation. The starting point is to consider

$$\frac{\mathrm{d}}{\mathrm{d}t}V(t) = \frac{1}{2}\int_0^{2\pi} \partial_t(\Delta_t^2)\,\mathrm{d}x = \int_0^{2\pi} \Delta_t\,(\partial_t \Delta_t)\,\mathrm{d}x. \tag{6.10}$$

With $\Delta_t = p_t - q_t$, as well as Eqs. (6.3) and (6.5), we have

$$\partial_t \Delta_t = \partial_t (p_t - q_t) = \partial_t p_t - \partial_t q_t \,,$$

resulting in the PDE

$$\begin{aligned}(\partial_t \Delta_t)(x) &= -\partial_x \left((\omega + Z(x)u(t))p_t(x)\right) + \partial_x \left(\omega q_t(x)\right) \\ &= -\omega(\partial_x \Delta_t) - \partial_x (Z(x)p_t(x))u(t)\,.\end{aligned}$$

Plugging the result in Eq. (6.10), we obtain

$$\frac{\mathrm{d}}{\mathrm{d}t} V(t) = -\omega \int_0^{2\pi} \Delta_t (\partial_x \Delta_t)\,\mathrm{d}x - \left(\int_0^{2\pi} \Delta_t \partial_x (Z p_t)\,\mathrm{d}x\right) u(t)\,.$$

Here the first term can be seen to be equal to $(\omega/2)\int_0^{2\pi} \partial_x (\Delta_t)^2 \,\mathrm{d}x = (\omega/2)[\Delta_t^2]_{t=0}^{t=2\pi} = 0$, where the vanishing is due to the 2π-periodicity of Δ_t^2. The final result is obtained via an integration by parts for the second term (described in more detail in Appendix E.4). ■

This result was independently obtained in B. Monga et al. (2018) where also a detailed derivation can be found. Given this result, it is immediate to apply the *population-level feedback law*

$$u(t) = -\left(\int_0^{2\pi} (\partial_x \Delta_t) Z p_t \,\mathrm{d}x\right), \tag{6.11}$$

which clearly results in $\frac{\mathrm{d}}{\mathrm{d}t} V(t) \leq 0$. To guarantee that $V(t) \to 0$, as $t \to \infty$, we need to study the set of solutions under which $\frac{\mathrm{d}}{\mathrm{d}t} V(t) \equiv 0$ similarly to the idea of LaSalle's invariance principle for the classical finite-dimensional case. Our main result on the convergence properties reads as follows.

Theorem 6.1. *Consider Eqs.* (6.3) *and* (6.5) *in closed loop with the control Eq.* (6.11)*. Suppose all Fourier coefficients of Z are non-zero, then, for all choices of p_0 and q_0, $p_t \to q_t$ as $t \to \infty$.*

Proof. It is clear that, with the given feedback law, we have $\dot{V} \leq 0$. To conclude the result, we need to additionally show that under the stated assumptions for the PRC Z, we can rule out the existence of solutions p_t and q_t for which $\Delta_t \neq 0$ but $\dot{V} \equiv 0 \Leftrightarrow u \equiv 0$. This would correspond to the system p_t evolving without external input, i.e., $\partial_t p_t = -\partial_x(\omega p_t)$, with the solution of the form $p_t(x) = p_0(x - \omega t)$. Similarly, we have $\Delta_t(x) = \Delta_0(x - \omega t)$, and thus $(\partial_x \Delta_t)(x) = (\partial_x \Delta_0)(x - \omega t)$. Under this assumption, we have

$$\int_0^{2\pi} (\partial_x \Delta_t) Z p_t \,\mathrm{d}x = \int_0^{2\pi} Z \Psi_t \,\mathrm{d}x = \langle Z, \Psi_t \rangle$$

with $\Psi_t : x \mapsto (\partial_x \Delta_0)(x - \omega t) p_0(x - \omega t)$. Suppose for the sake of contradiction that $\Delta_0 \neq 0$. As Δ_0 is a difference of two probability densities, it furthermore cannot be a constant function, so that we can conclude $\partial_x \Delta_0 \neq 0$. As the product of two non-zero functions, of which one is positive, we can conclude that Ψ_0 is non-zero and thus has a non-vanishing sequence of Fourier coefficients (ψ_k). Now, by Parseval's theorem, we have

$$u(t) = -\int_0^{2\pi} Z \Psi_t \,\mathrm{d}x = -\langle Z, \Psi_t \rangle = -\sum_{k=-\infty}^{\infty} c_k \psi_k \,\mathrm{e}^{-\mathrm{i}k\omega t},$$

where $\psi_k \mathrm{e}^{-\mathrm{i}k\omega t}$ is the kth Fourier coefficient of Ψ_t. We can view the sequence $(c_{-k}\psi_{-k})$ as the Fourier coefficients of the function $u_\omega : t \mapsto u(t/\omega)$. Since this sequence is non-zero by the assumption that $c_k \neq 0$, $\forall k \in \mathbb{Z}$ and the non-vanishing of the sequence (ψ_k), we can conclude that u_ω and thus u is not identically zero, yielding the contradiction. ∎

Theorem 6.1 relies on a PRC with non-vanishing Fourier coefficients. This condition is rarely met in a biological context where the shape of the PRC is determined by rather smooth dynamics in the biological system. Examples of PRCs from circadian clock and neuron activity are shown in Figure 6.2. The absolute value of the Fourier coefficients decreases rapidly with the 5th coefficient being practically zero in all cases. Based on Theorem 6.1 the feedback Eq. (6.11) would not result in convergence of p_t to q_t in these examples. Thus, we considered variants of the cost functional Eq. (6.8) in which we are not interested in the L_2-norm of the difference Δ_t but rather some projection $P(\Delta_t)$, where $P : L_2 \to L_2$ is an orthogonal projection onto a finite-dimensional subspace of L_2. A particularly important special case is given by setting

$$P = P_N : p_t \mapsto \frac{1}{2\pi} \sum_{k=-N}^{N} \mathrm{e}^{\mathrm{i}k(\cdot)} \langle \mathrm{e}^{-\mathrm{i}k(\cdot)}, p_t \rangle \,, \tag{6.12}$$

which would yield a case where we are only interested in the tracking for the first N Fourier coefficients of p_t and q_t as $t \to \infty$. Thus, in the following we will consider the more general cost functional

$$V(t) = \frac{1}{2} \langle P\Delta_t, P\Delta_t \rangle. \tag{6.13}$$

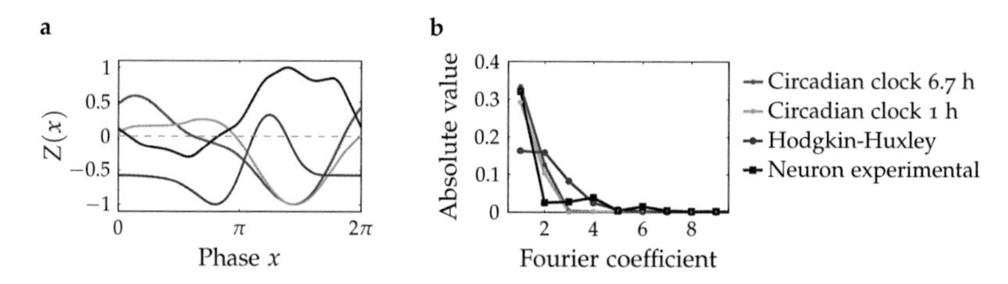

Figure 6.2. Circular moments of real-world PRCs decay rapidly. (**a**) PRCs of the circadian clock in response to a 6.7h and 1h light impulse (green). Curve obtained experimentally and reported in St Hilaire et al. (2012). PRCs of neurons from Hodgkin-Huxley model (Ashwin et al. 2016) and from experiments (Stiefel et al. 2008) (blue). (**b**) Absolute values of the Fourier coefficients of the PRCs. © 2019 IEEE.

For this modified case, we have the following main result.

Theorem 6.2. *Let $\nu_Z \in \mathbb{N}$ be such that $c_k \neq 0$ for $|k| \leq \nu_Z$. Then, Eqs. (6.3) and (6.5) in closed loop with the feedback law*

$$u(t) = -\left(\int_0^{2\pi} P_{\nu_Z}(\partial_x \Delta_t) Z p_t \,\mathrm{d}x \right)$$

will result in the convergence of the first ν_Z moments of p to the first ν_Z moments of the target distribution q.

Proof. A direct computation along the lines of the previous case shows that

$$\frac{\mathrm{d}}{\mathrm{d}t}V(t) = \left(\int_0^{2\pi} P(\partial_x \Delta_t) Z p_t \,\mathrm{d}x\right) u(t)\,.$$

With $P = P_{\nu_Z}$ and the above feedback law, we clearly have $\frac{\mathrm{d}}{\mathrm{d}t}V(t) \leq 0$. It is left to show that under the assumption on Z, there can be no non-trivial configuration of p and q resulting in $\frac{\mathrm{d}}{\mathrm{d}t}V \equiv 0 \Leftrightarrow u \equiv 0$. Again, this would correspond to the system p_t evolving without external input, i.e., $\partial_t p_t = -\partial_x(\omega p_t)$, the solution of which is of the form $p_t(x) = p_0(x - \omega t)$, and similarly, $\Delta_t(x) = \Delta_0(x - \omega t)$. Let d_k denote the Fourier coefficients of Δ_0. Then

$$(\partial_x \Delta_t)(x) = \frac{1}{2\pi} \sum_{k=-\infty}^{\infty} (\mathrm{i}\, k d_k\, \mathrm{e}^{-\mathrm{i}\, k\omega t})\, \mathrm{e}^{\mathrm{i}\, kx}$$

and thus

$$(P_{\nu_Z}(\partial_x \Delta_t))(x) = \frac{1}{2\pi} \sum_{k=-\nu_Z}^{\nu_Z} (\mathrm{i}\, k d_k)\, \mathrm{e}^{\mathrm{i}\, k(x-\omega t)}\,.$$

Thus, the integral under consideration can again be written as

$$u(t) = -\int_0^{2\pi} Z \Psi_t \,\mathrm{d}x\,, \tag{6.14}$$

where

$$\Psi_t : x \mapsto (P_{\nu_Z}(\partial_x \Delta_0))(x - \omega t) p_0(x - \omega t)\,. \tag{6.15}$$

The map Ψ_0 is given by

$$\psi_k = \sum_{\ell=-\nu_Z}^{\nu_Z} \mathrm{i}\, \ell d_\ell m_{k-\ell}\,, \quad k \leq \nu_Z \tag{6.16}$$

and the kth Fourier coefficient of Ψ_t is given by $\psi_k\, \mathrm{e}^{-\mathrm{i}\, k\omega t}$. Now the mapping $(d_k)_{|k|\leq \nu_Z} \mapsto (\psi_k)_{|k|\leq \nu_Z}$ is a linear mapping involving the Toeplitz matrix T_{ν_Z} generated by the Fourier coefficients m_k of p_0 (Grenander and Szegő 1958)

$$T_{\nu_Z} = \begin{bmatrix} m_0 & m_1 & \dots & m_{2\nu_Z} \\ m_{-1} & m_0 & \ddots & \vdots \\ \vdots & \ddots & \ddots & m_1 \\ m_{-2\nu_Z} & \dots & m_{-1} & m_0 \end{bmatrix}. \tag{6.17}$$

Since p_0 is positive, this Toeplitz matrix is positive definite so that we can conclude that $(\psi_k)_{|k|\leq \nu_Z}$ is non-zero. By the same line of argument as before, the claim follows. ■

6.2.2 Properties of the ensemble controller

We will now discuss some properties of the proposed ensemble control approach. First, we discuss the results in light of real biological systems and their phase response curves. Second, we revise the controllability results for the relevant control goal of synchronizing a populations.

Real world PRCs

A commonly observed phenomenon is that smoothness of a function correlates with the rate of decay of its Fourier coefficients. The dependence of the statement in Theorem 6.2 on the coefficients in the phase response curve can be interpreted in light of the smoothness of the functions. The degree of controllability of the distribution of agents on the unit circle is determined by the ruggedness of the PRC. The distribution of agents can be controlled to the same ruggedness as the ruggedness of the PRC by which the input affects the system. Given the well-known relation on rate of decay of Fourier coefficients for continuous periodic functions and their smoothness, our results practically restrict the controllability of oscillators in an ensemble control framework to only few circular moments (Polya and Wiener 1942). However, a fast rate of decay of Fourier coefficients correlates with the rate of convergence of the Fourier transform to the actual function, such that smooth target distribution might be achieved with high precision.

Synchronization of the population

Synchronizing a population of agents on their limit cycle is a common fundamental control goal which is naturally included in the above described setup. A synchronized population is commonly achieved by sending the length of the first moment to one, $|m_1| \to 1$, cf. Kuritz et al. 2018a. Based on Theorem 6.2 we arrive at the following result: Suppose $\nu_Z \geq 1$ and a target distribution characterized by $|\alpha_1| = 1$, then the feedback law

$$u(t) = -\left(\int_0^{2\pi} P_1(\partial_x \Delta_t) Z p_t \, \mathrm{d}\, x\right)$$

will result in synchronization of the population with $|m_1| \to 1$. Thus, in this special case where one tries to achieve a synchronized population, it is sufficient to control only the first moment which is possible if the first moment of the PRC is non-zero.

6.3 Computational studies

We illustrate the performance and also limitations of our controller in three different scenarios. To this end artificial phase response curves, as well as target distributions were generated and the number of moments under control was varied. We will set the following variables in the different scenarios such that all relevant consequence of Theorem 6.2 emerge:

- ν_q: index of the largest non-zero Fourier coefficient of the target distribution.
- ν_Z: index of the largest non-zero Fourier coefficient of the phase response curve.
- ν_u: number of Fourier coefficient used in the projection P for calclulating the input u.

The scenarios are summarize in Table 6.1.

The PRCs were constructed to have identical Fourier coefficients for all moments such that $Z(x) = \sum_{k=-\nu_Z}^{\nu_Z} 1\, \mathrm{e}^{\mathrm{i}kx}$. The circular moments of the target distributions were chosen to be

Table 6.1. Simulation scenarios

Scenario	Moments	Performance
1	$\nu_Z = \nu_u = \nu_q$	Higher moments evolve.
2	$\nu_Z = \nu_u > \nu_q$	Distributions are well aligned.
3	$\nu_Z < \nu_u = \nu_q$	Target moments can not be achieved.

evenly distributed at equal distance from the origin

$$q_0(x) = \frac{1}{2\pi} \sum_{k=-\infty}^{\infty} \alpha_k^0 \, e^{ikx} \,,$$
$$\text{with} \quad a_k^0 = \begin{cases} 1 & \text{for } |k| = 0 \\ \frac{1}{8} \, e^{i\left(\frac{\pi}{8} + \frac{k}{2\pi\nu_q}\right)} & \text{for } |k| = 1, \ldots, \nu_q \\ 0 & \text{for } |k| > \nu_q \end{cases} . \tag{6.18}$$

The system in scenario 1 is initialized with a distribution with all moments equal to zero except $m_3 = 0.05$. In all other scenarios, the system is initialized with a uniform distribution $p_0(x) = (2\pi)^{-1}$.

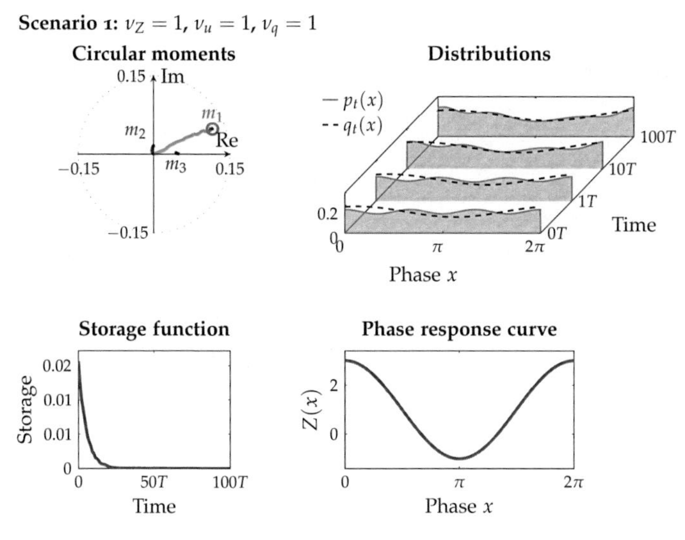

Figure 6.3. Scenario 1 with $\nu_Z = 1$, $\nu_u = 1$ and $\nu_q = 1$. © 2019 IEEE.

We will briefly summarize the most important observations in general and individually for the three scenarios. In any case, moments larger than ν_Z cannot be controlled and might evolve as side effect of the control of the moments smaller or equal ν_Z. This effect can be seen in Scenario 1 where the second moment deviates significantly from zero and the third moment stays close to the value where it was initialized (Figure 6.3). The effect can be reduced by controlling more moments than present in the target distribution as shown in Scenario 2,

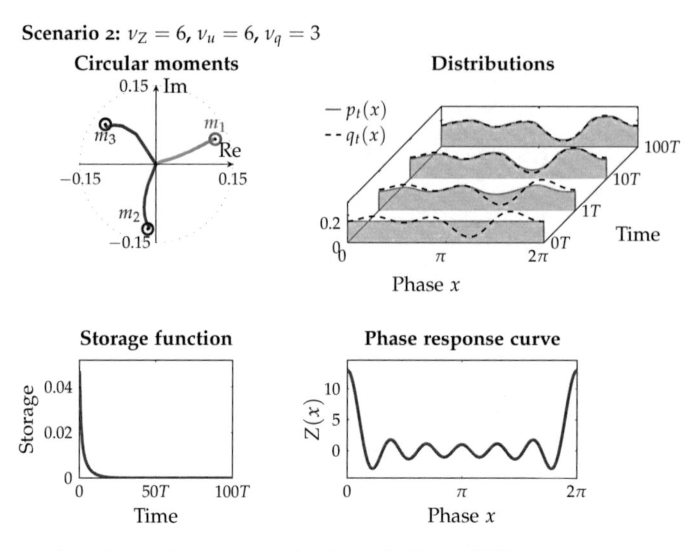

Figure 6.4. Scenario 2 with $\nu_Z = 6$, $\nu_u = 6$ and $\nu_q = 3$. © 2019 IEEE.

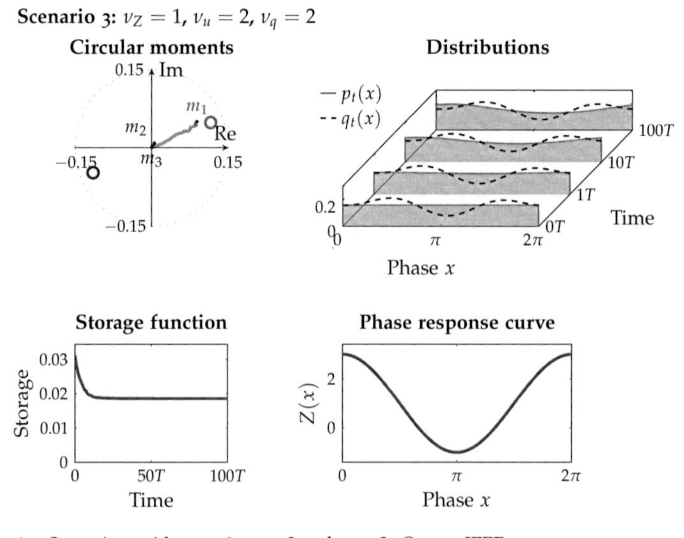

Figure 6.5. Scenario 3 with $\nu_Z = 1$, $\nu_u = 2$ and $\nu_q = 2$. © 2019 IEEE.

where ν_Z is much larger than ν_q (Figure 6.4). Finally, implications of Theorem 6.2 are clearly demonstrated in Scenario 3 (Figure 6.5). Only the first moment of the PRC is non-zero and both non-zero moments of the target distribution were used in the controller. The input converges to zero, however, as a consequence of Theorem 6.2, invariant distributions other than those with $P(\Delta_t) = 0$ exist, leading to oscillations with a distribution far off the desired one and neither of the moments converges to the corresponding one of the target distribution (Figure 6.5).

Finally, we study the special case of synchronizing a population of oscillators by tracking a target distribution with the first moment equal to one, $|\alpha_1| = 1$. The first moment of p tends towards the unit circle as the cells approaches a synchronized population (Figure 6.6). A synchronized population corresponds to a Dirac delta distribution which by definition has the length of all moments equal to one. As a consequence of synchronizing the population by controlling the first moment, all moments approach the unit circle. A sufficient condition to achieve synchrony is that only the first moment of the PRC is non-zero. This result is in line with the previous results in Chapter 5 on cell cycle synchronization where only the length of the first moment was considered in the controller design.

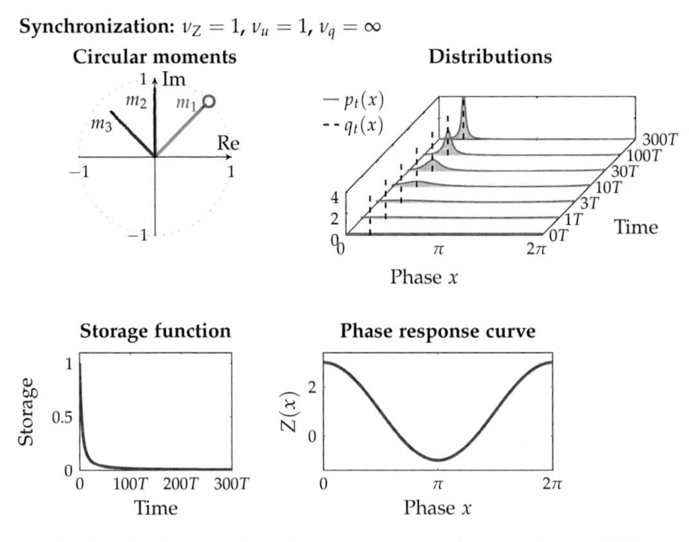

Figure 6.6. Synchronization scenario with $\nu_Z = 1$, $\nu_q = \infty$ and $\nu_u = 1$. © 2019 IEEE.

6.4 Summary and discussion

Summary

In this chapter, we investigated the control problem of achieving a desired distribution of cellular oscillators on the periodic orbit of their limit cycle. Using the reduced phase model representation of oscillatory systems a cost functional was formulated that promoted a population-level feedback law as the solution. A LaSalle-like argument was used to derive novel analytical conditions for the ability to control phase distributions in terms of the Fourier coefficients of the phase response curve. Our main result states that full control to any desired distribution is only possible if all moments of the PRC are non-zero. Moreover, we derived a condition for convergence of the first N moments of the population to those of the target distribution, which requires that only the first N moments of the phase response curve are non-zero. In this light, we also discussed how the task of achieving synchrony is a special case where it is sufficient that the first moment of the PRC is non-zero as a synchronized population is specified already by its first moment (or any other) having length equal to one.

Discussion and Outlook

The feedback control is applicable to many systems given the input response occurs rapidly and the communication between individual oscillators is weak. The symptoms in Parkinson's disease, for example, are associated with elevated synchrony of neurons, and the reduction of this synchrony by deep brain stimulation is correlated to the alleviation of the symptoms. A compelling extension of this work is to consider networks of coupled oscillators, whose interactions are characterized by a coupling function acting between each pair of oscillators. In the future, it might be interesting to extend the theory to noisy systems which present more realistic models with regard to real applications. The noise might even have a positive effect on the control problem as it pushes the moments to the origin so that higher moments in the population which cannot be controlled, due to zero moments of a realistic PRC, might thereby vanish.

Chapter 7

Conclusions

Single-cell data is rich in information. This is particularly the case if heterogeneity in a population establishes by means of a biological process over which cells in the population spread. Recent development in single-cell technologies brought great opportunities and challenges for the analysis of biological processes. These developments gave rise to the novel and highly active research field of trajectory inference within the field of systems biology and single cell analysis. The explosion in single-cell studies in recent years was accompanied by the development of a plethora of trajectory inference algorithms. We took the pseudotemporal ordering provided by trajectory inference algorithms as a starting point from which we developed our results.

Furthermore, ever growing possibilities in manipulating biological systems, e.g. with optogenetic tools, paved the way for new approaches to control these systems. In combination with advanced analysis methods population-level feedback algorithms may provide broadcast inputs to, for example, facilitate rigorously controlled experimental conditions for basic research or reconstitute healthy cell states in the treatment of diseases.

The theoretical results described in the thesis led to readily applicable algorithms which we, whenever possible, applied to real experimental data to solve problems motivated by a biological question. Theoretical results were furthermore illustrated by various examples. The results in this thesis were structured in two parts. Part I was devoted to analysis methods for biological processes in single-cell data. Part II presented novel ensemble control algorithms for cellular oscillators. Below, we summarize the results derived in the individual chapters in greater detail and discuss general aspects. Furthermore, we point to possible directions for future research as an outlook.

7.1 Summary and discussion

Part I Analysis

Chapter 2 of this thesis was concerned with studying a shortcoming of trajectory inference which is the arbitrariness of the pseudotime. The MAPiT algorithm, which we presented in Chapter 2, solves this problem by transforming the pseudotime scale to a real scale. We demonstrated the method for transformation to a real-time scale in cell cycle analysis and a spatial scale in experiments with multicellular tumour spheroids. An important feature of MAPiT is its independence of the actual pseudotime scale. We demonstrated the robustness of MAPiT to changing pseudotime scales, by comparing its performance to recover the spatial position using different pseudotime algorithms, markers and spheroid sizes. By recovering temporal and spatial information MAPiT enables dynamic analysis of complex biological systems in an high-throughput and high-content fashion. We demonstrated the capability of MAPiT towards high-throughput, by examining the composition of spheroids at different diameters. This experiment provided conserved molecular profiles throughout different spheroid sizes, indicating a dependence of these markers solely on the distances from the surface.

Concluding, our results in Chapter 2 provide a solution to a fundamental problem, namely the transformation of pseudotime to real-time or the true scale of a biological process. We presented with MAPiT a robust and universal tool to recover temporal or spatial cellular trajectories from high-dimensional single-cell experiments, enabling analysis of process dynamics in high-throughput and high-content setups.

Chapter 3 employed ergodic principles to address the cell cycle problem from Chapter 2 in a dynamical systems point of view. Ergodic theory in principle states that the number of cells in a particular stage of a process is related to the average transit time through that stage. By applying this concept to the description of the progression of a single cell through the cell cycle by a stochastic differential equation we established a connection between pseudotemporal ordering and age-structured population models. In this context we provided the mapping from pseudotime to cell age in three approaches where a cell proceeds through the cell cycle (1) deterministically without noise, (2) with addition of a Wiener process representing intrinsic noise or (3) deterministically with distributed progression rate representing extrinsic noise. In the first case where the noise was omitted, we showed analytically, that the derived transformation results in the steady state age-structure of age-structured cell population models and thus coincides with the results from Chapter 2. In noisy progression settings the one-to-one mapping from pseudotime to cell age becomes a convolution by the solution of a PDE. The stochastic differential equation in the second approach with intrinsic noise, resulted in a Fokker-Planck-type PDE for the transformation and we presented a novel identification algorithm for state-dependent progression and noise functions. The third approach with extrinsic noise is characterized by a distributed cell cycle progression rate which we deduced from the distribution of total cell cycle lengths in the population. Taken together, noise consideration resulted in a transformation to a true age scale while the transformation without noise results in a transformation to a cell cycle stage scale which locally equals time.

Concluding, our results in Chapter 3 established a connection between cell cycle analysis in pseudotime and age-structured population models. The results form the basis for various research directions in the field of single cell analysis which we for example exploited in Chapters 4 and 5.

Chapter 4 is dedicated to an application of the concepts and methods developed in Chapters 2 and 3 and provided extension to non-stationary distributions. The methods in the previous chapters were based on the steady state assumption which restricts the application of these methods to untreated populations. A compelling extension of these approaches are studies where the stationary assumption is not satisfied which, for example, emerges in response to drug treatment. A treatment of the population in steady state results in a change of cell cycle progression which manifests in altered cell cycle dependent distributions. These changing distributions reflect a time- and position-dependent change in cell cycle progression, and we provided an efficient learning algorithm for the inference thereof. To achieve this, we derived the sensitivity system of parameters in the progression function for age-structured type population models. We showcased the capacity of the method to recover unknown parameters with simulated data. When applied to two experimental data sets, our method uncovered new insights. First, it recognized in the estimated speed change function that Nocodazole induced cell cycle arrest in M-phase. Second, the effects of batch culture cause significant changes in cell cycle progression.

Concluding, our results in Chapter 4 provided a computational framework that allows efficient inference of changes in cell cycle progression. We thereby contribute to current efforts in the field of single cell analysis to trace the rewiring of regulatory networks in response to treatments in different cellular contexts, e.g. cell cycle stages (Wijst et al. 2020; Holland et al. 2020).

Part II Control

In the second part of this thesis, we turned our attention from analyzing a biological process in a heterogeneous population to controlling it, in the sense, that we wanted to achieve a desired distribution of cellular oscillators on their periodic orbit. Motivated by restrictions brought into the problem by the nature of biological data and the ability to manipulate cell populations, two ensemble control formulations were sought.

Chapter 5 introduced the ensemble control problem for cell cycle synchronization. Building upon the theory of reduced phase models for oscillatory systems with a limit cycle, we arrived at an age-structured population model similar to the ones of Chapters 3 and 4. We formulated the problem of cell cycle synchronization for this model as sending the norm of the first circular moment to one. By a transformation of the state, the model was brought to a favorable form for controller design in an input-/output setting. We proved stability of our derived controller by showing passivity of the overall system. Convergence analysis was carried out by LaSalle-like arguments. The analysis of invariant sets then led to the controllability condition which calls for the first moment of the phase response curve to be non-zero. Theoretical results were illustrated by simulation studies with the reduced phase model. In addition, simulation studies with a realistic individual-based simulation framework provided suitable parameter ranges in which the population synchronizes despite the naturally heterogeneity. Examining the performance of the controller in this realistic individual-based framework provided a first proof-of-concept. This is crucial due to the difficulty in verifying the assumptions imposed by the reduced phase model approach, such as, the neighborhood around the limit cycle in which the theory holds.

Concluding, our results in Chapter 5 provided a readily applicable ensemble control setup for cell cycle synchronization. Several experimental methods rely on denoising the population from cell cycle effects. Our ensemble controller contributes to this challenge as it allows to continuously maintain a synchronized population with minor side-effects as compared to standard approaches.

Chapter 6 is dedicated to the control problem of achieving a desired distribution of cellular oscillators on the periodic orbit of their limit cycle. Our results were motivated by diseases including cancer, Parkinson's disease and heart diseases, which are accompanied by loss or malfunction of regulatory mechanism of an oscillatory system. Successful treatment of these diseases might involve recovering the healthy behavior of the oscillators in the system, i.e., achieving synchrony or a desired distribution of the oscillators on their periodic orbit. As done in Chapter 5, we used the reduced phase model representation of oscillatory systems to formulate a cost functional that promoted a population-level feedback law as the solution. A LaSalle-like argument was used to derive novel analytical conditions for the ability to control phase distributions in terms of the Fourier coefficients of the phase response curve. Our main result in Theorem 6.1 states that full control to any desired distribution is only possible if all moments of the PRC are non-zero. Moreover, we stated in Theorem 6.2 a condition for convergence of the first N moments of the population to those of the target distribution, which requires that the first N moments of the phase response curve are non-zero. A commonly observed phenomenon is that smoothness of a function correlates with the rate of decay of its Fourier coefficients. In this light, we interpreted our controllability condition in the sense that the ruggedness of the PRC specifies the reachable amount of ruggedness in the target distribution. Simulation studies with varying PRCs and target distributions illustrated these observations.

Concluding, our results in Chapter 6 provided a systems theoretic approach to the ensemble control problem for cellular oscillators and novel necessary and sufficient conditions for the control of phase distributions in terms of the Fourier coefficients of the phase response curve. The presented controller is applicable to many systems, i.e. phase shift for circadian clock or desynchronizing of spiking neurons in Parkinson's disease.

Last but not least, it is essential to test and evaluate the ensemble control algorithms from Chapters 5 and 6 thoroughly in experiments and practical applications. Not only is this necessary in order to prove the practicality and reliability of novel algorithms, but also have challenging applications always been a driving force for the theoretic progress.

7.2 Outlook

The endeavor of developing algorithms for single cell analysis and control for cellular ensembles has just started. It is less than a decade ago that the first trajectory inference algorithms have been presented for analysing structured single-cell data (Bendall et al. 2014; Trapnell et al. 2014). Development of these algorithms was driven by new experimental technologies like CyTOF or single-cell RNA-sequencing, which called for advanced analysis algorithms. Promoted by early successes and the enormous potential of single cell analysis methods, experimental technologies and computational frameworks developed rapidly.

For example, pseudotime algorithms were recently further developed to robustly recognise also branching processes in differentiation pathways. The methods described in Chapters 2 and 3 could be used to study the dynamics in individual branches. For instance, MAPiT could be applied to all paths from the root to the end points on the respective branches treating the flow of cells into other branches as sinks. However, MAPiT relies on the real-time distribution which might not be directly available in such settings.

A further extension of the results from this thesis would be to relax the assumption on the homogeneity of the population, in the sense that all cells at one stage of the process behave in the same way. This assumption might not be met if a treatment would affect the progression rate of a fraction of the population while the rest of the population is unaffected. Successful real-time analysis would involve the identification of (1) the size and (2) the progression rate of the subpopulations, which could be achieved by a mixture model approach.

Furthermore, assuming that the cells in a cell population progress and react independently is a simplifying assumption neglecting couplings between the cells which may form directly or indirectly through a common medium. Extending the presented theory and algorithms to populations with couplings brings fascinating opportunities, such as studying cell-to-cell communication in snapshot data.

It is of immediate interest to apply our methods to the analysis of differentiation pathways, the most common application for single cell analysis. Achieving the transformation of pseudotime to the real-time axis however is more challenging for cell differentiation. In cell differentiation pathways information on the real-time distribution is not available and additional information about cell division and death rates would be needed for the application of ergodic principles for computing the velocity function. This is due to the non-identifiability of velocity and combined entry and exit rates at the same time as we discussed in Section 2.5.

We envision that the integration of additional data and models with pseudotime algorithms might compensate for a lack in prior knowledge or unmet assumptions when recovering real-time dynamics from single-cell data. We presented one such example in Chapter 4 where a treatment caused time- and cell cycle stage-dependent change in progression. Similarly, a cell population will not fulfill the steady-state assumption if cell death or cell division rates change along the process and with time. These events however may be recorded by suitable markers which can be learned from large scale gene expressions (Szalai et al. 2019). Similarly, phosphohistone-3 may directly capture cells in M-phase and cleaved Caspase-3 may serve as indicator for apoptosis.

These readings could then be integrated as marker-dependent rates in the reduce model, resulting in a PDE model with dynamics

$$\partial_t n(x,t) + \partial_x \left(f(x,t)\, n(x,t) \right) = \left(\alpha(y) - \beta(y) \right) n(x,t) \,.$$

The functions $\alpha(y)$ and $\beta(y)$ represent marker-dependent division and death rates and could be inferred directly from the measurements.

The presented concepts could furthermore contribute to efforts towards building genome scale cell signalling networks. Such models have great potential in the area of precision medicine where a therapy design might be guided by the prediction of drug responses in individual patients (Saez-Rodriguez and Blüthgen 2020). Building and training of these dynamic models requires large amounts of data and could benefit substantially from single cell snapshot data in combination with real-time analysis. The method that we presented in Chapter 4 where we estimated the change in cell cycle progression is capable of generating real-time cell cycle stage-dependent molecular dynamics in response to drug treatment at omics scale. The molecular dynamics provide scope to study rewiring of signaling networks in response to drug treatment and to identify sensitive nodes in drug response or during the evolution of diseases.

These possible directions for future research address individual aspects which allow to relax certain assumptions in single cell analysis of biological processes. A next step would naturally constitute the combination of several extensions to the present theory. Things might get messy and we believe that it might be worthwhile to make a step back and elementarily formulate the model of the process. This would allow to rigorously characterize fundamental systems theoretic properties such as observability and controllability conditions for biological processes in single cell analysis.

In Chapter 3, we developed the transformation from pseudotime to cell age for the specific problem of stochastic cell cycle progression (Kuritz et al. 2017). Transferring the results described in Chapters 5 and 6 in this thesis into the stochastic framework thus is a consequent continuation of the present work.

Another compelling extension of our results on ensemble control for cellular oscillators is to consider networks of coupled oscillators, whose interactions are characterized by a coupling function acting between each pair of oscillators.

Furthermore, combining learning and adaptive control approaches with our ensemble control algorithms may provide scope to identify functions that govern the behavior of the system, like the PRCs or coupling functions, online. In this way, the control algorithms could adapt to changing dynamics caused, for example, by changing environments or changes in the health state of a patient.

This brings us to our long-term vision where we imagine a combination of (1) identifying biological processes and learning changes in response to treatments with (2) control of ensembles in biomedical processes to achieve a desired behavior. Recent development in the wearables market, computing power and machine learning algorithms provides the opportunities to combine the approaches presented in this thesis. New wearables technology is capable of measuring a multitude of parameters in a person, ranging from body temperature and heart rate to blood glucose levels or cerebral nerve activity. We thus aim at changing healthcare medicine with our tools. We imagine a combination of drug and disease modeling and simulation with advanced data analytics and control theory to pave the way for online system analysis and therapy design to enhance precision medicine and accelerate drug discovery.

Appendices

Appendix A

Technical background

This chapter presents the description of the theoretic framework underlying the ensemble control approaches dealt with in this thesis. Appendix A.1 is dedicated to nonlinear systems and control theory. Appendix A.2 presents the basic concept of reduced phase models for oscillatory systems. Finally, Appendix A.3 introduces Fourier coefficients and circular moments for oscillatory systems.

A.1 Dynamical systems and control theory

We review here some basic concepts from dynamical systems, stability and feedback control. The presentation of this section follows the textbooks Desoer and Vidyasagar (1975) and Khalil (2002).

A.1.1 Dynamical system with state space

In general, a continuous *control system* is a tuple $(\mathcal{X}, \mathcal{U}, \mathcal{X}_0, f)$, where $\mathcal{X} \subset \mathbb{R}^n$ is a non-empty set called the state space, the input set $\mathcal{U} \subset \mathbb{R}^l$ is a non-empty set containing the origin and $\mathcal{X}_0 \subset \mathcal{X}$ is the set of initial conditions. The function $f\colon \mathcal{X} \times \mathcal{U} \mapsto \mathcal{X}$ is called the control vector field. For a given initial condition $x(0) \in \mathcal{X}_0$ and control input function $u\colon \mathbb{R}_{\geq 0} \mapsto \mathcal{U}$, the map $x\colon \mathbb{R}_{\geq 0} \mapsto \mathcal{X}$ satisfying the ordinary differential equation

$$\dot{x}(t) = f(x(t), u(t)), \quad t \in \mathbb{R}_{\geq 0} \tag{A.1}$$

is called a solution of the continuous time control system. The system Eq. (A.1) is time-invariant. If f depends explicitly on the time t, we say that the control system is time-variant. If f does not depend on u, then we speak of a *dynamical system* instead of a control system.

An important concept in control theory is that of an *equilibrium point*. A point $x^\star \in \mathcal{X}$ is an equilibrium point if the constant function $x(t) = x^\star, \forall t \in \mathbb{R}_{\geq 0}$ is a solution of Eq. (A.1). For an unforced continuous time control system it is clear that $x^\star$ is an equilibrium point if $0 = f(x^\star, 0)$.

With the definitions of an equilibrium point at hand we can now introduce the notions of invariance, stability and attractivity. Stability is commonly characterized in the sense of Lyapunov (Krasovskii 1963; Khalil 2002), LaSalle's invariance principle (La Salle 1966), or by passivity-based results in input-/output frameworks (Desoer and Vidyasagar 1975). Loosely speaking, an equilibrium point is stable if all solutions starting at nearby points stay nearby; otherwise, it is unstable. Furthermore, it is asymptotically stable if all solutions starting at nearby points not only stay nearby, but also tend to the equilibrium point as time approaches infinity. A set is invariant, if we stay inside the set whenever we start inside the set.

Consider a time-invariant dynamical system with state space $\mathcal{X}$ and let $\mathcal{S} \subset \mathcal{X}$. The set $\mathcal{S}$ is *positively invariant* under the dynamical system if each solution $x(t)$ with initial condition in $\mathcal{S}$ remains in $\mathcal{S}$ for all times.

$$x(0) \in \mathcal{S} \Rightarrow x(t) \in \mathcal{S}, \forall t \geq 0 .$$

Furthermore, we say that $x(t)$ approaches a set $\mathcal{S}$ as t approaches infinity, if for every neighborhood $\mathcal{W}$ of $\mathcal{S}$, there exists a time $t_0 > 0$ such that $x(t)$ takes values in $\mathcal{W}$ for all subsequent times $t \geq t_0$. We thus arrive at the following definition of stability:

Definition A.1 (Stability). *Consider a time-invariant dynamical system. A set $\mathcal{S}$ is said to be*

1. stable *if, for any neighborhood $\mathcal{Y}$ of $\mathcal{S}$, there exists a neighborhood $\mathcal{W}$ of $\mathcal{S}$ such that every solution of Eq.* (A.1) *with initial condition in $\mathcal{W}$ remains in $\mathcal{Y}$ for all subsequent times;*
2. unstable *if it is not stable;*
3. locally attractive *if there exists a neighborhood $\mathcal{W}$ of $\mathcal{S}$ such that every trajectory with initial condition in $\mathcal{W}$ approaches the set $\mathcal{S}$;*
4. locally asymptotically stable *if it is stable and locally attractive.*

Furthermore, the set $\mathcal{S}$ is globally attractive *if every evolution of the dynamical system approaches it and it is* globally asymptotically stable *if it is stable and globally attractive.*

We will now briefly state the invariance principle for continuous time systems, known as *LaSalle's theorem*. Consider a continuous time dynamical system with state space $\mathcal{X}$ and control vector field f. For a function $V\colon \mathcal{X} \mapsto \mathbb{R}$, the directional derivative along the solutions of the dynamical system Eq. (A.1) is defined as

$$\dot{V} = \frac{\partial V}{\partial x} f(x) . \tag{A.2}$$

This directional derivative is the time derivative along the a specific solution $x(t)$ of the dynamical system $\dot{x} = f(x)$ and can be computed as $\dot{V} = \frac{\mathrm{d}}{\mathrm{d}t} V(x(t))$. Given a set $\mathcal{D}$, if $\dot{V}(x) \leq 0,\ \forall x \in \mathcal{D}$, then the function $V(x)$ is *non-increasing* on a set $\mathcal{D} \subset \mathcal{X}$ along the solution of the dynamical system. With these definitions at hand, we can now state the invariance principle for continuous time systems:

Theorem A.1 (LaSalle). *Let $\mathcal{S} \subset \mathcal{X}$ be a compact set that is positively invariant with respect to the dynamical system. Suppose there exists a function $V\colon \mathcal{D} \mapsto \mathbb{R}$ that is continuously differentiable and non-increasing along the trajectories of the dynamical system, i.e. $V(x) \leq 0,\ \forall x \in \mathcal{S}$. Let $\mathcal{E}$ be the set of all points in $\mathcal{S}$ where $\dot{V}(x) = 0$ and let $\mathcal{M}$ be the largest positively invariant set contained in $\mathcal{E}$. Then every solution starting in $\mathcal{S}$ approaches $\mathcal{M}$ as $t \to \infty$.*

The assumptions imposed by LaSalle's theorem may in some cases be to restrictive, such that other invariance-like results may be required. Barbalat's lemma is in particular useful for analyzing the asymptotic behavior of continuous time dynamical control systems:

Lemma A.1 (Barbalat's lemma). *Let $\psi\colon \mathbb{R} \mapsto \mathbb{R}$ be a uniformly continuous function on $[0, \infty)$. Suppose that $\lim_{t\to\infty} \int_0^t \psi(\tau)\,\mathrm{d}\,\tau$ exists and is finite. Then $\psi(t) \to 0$ as $t \to \infty$.*

A.1.2 Input-/output mapping and control approach

We turn our attention now to systems with input and outputs. A continuous time control system with input and outputs is represented by

$$\dot{x}(t) = f(x(t), u(t)), \quad y(t) = h(x(t), u(t)), \qquad t \in \mathbb{R}_{\geq 0}\,, \tag{A.3}$$

with the addition outputs $y \in \mathcal{Y} \subset \mathbb{R}^q$. In an input-/output framework however the model description is relaxed to the input-/output relation by

$$y = Hu \tag{A.4}$$

where H is some mapping or operator that specifies y in terms of u. Thus, we note that a system can now be recast as an input-/output mapping of an input signal u to an output signal y. Following the formal framework treated in Desoer and Vidyasagar (1975), let $x : \mathbb{R}_{\geq 0} \to \mathbb{R}$ be a scalar function of time and

$$x_T = \begin{cases} x(t), & t \leq T \\ 0, & t > T \end{cases} \tag{A.5}$$

the T-truncated signal. Given the L^2 inner product

$$\langle x, y \rangle = \int_0^\infty x(t) y(t)\, \mathrm{d}t\,, \tag{A.6}$$

we let

$$\mathcal{L}_e \triangleq \{x\colon \forall T \in \mathbb{R}_{\geq 0}\,,\ \langle x_T, x_T \rangle < \infty\} \tag{A.7}$$

be the space of signals x with the property that all truncations have finite L^2-norm and

$$\mathcal{L} \triangleq \{x\colon\ \langle x, x \rangle < \infty\} \tag{A.8}$$

the space of signals for which this holds for the complete signal.

We now define the mapping

$$\begin{aligned} H_1 :\quad \mathcal{L}_e &\to \mathcal{L}_e\,, \\ u &\mapsto y \end{aligned} \tag{A.9}$$

which takes an arbitrary input signal $u \in \mathcal{L}_e$ and returns the output signal $y \in \mathcal{L}_e$, depending on the initial distribution and its evolution dynamics.

Given this approach, the passivity of such a system can be studied using the classical input-/output framework treated in Desoer and Vidyasagar (1975), avoiding the difficulties of formulating a proper state space for defining a storage function. With the mapping H_1 defined by the system dynamics, the goal is to apply an output feedback approach as depicted in Figure A.1.

A.2 Weakly connected oscillators

In the following we review the basic concept of reduced phase models and phase response curves briefly and refer the interested reader to the book Hoppensteadt and Eugene M.

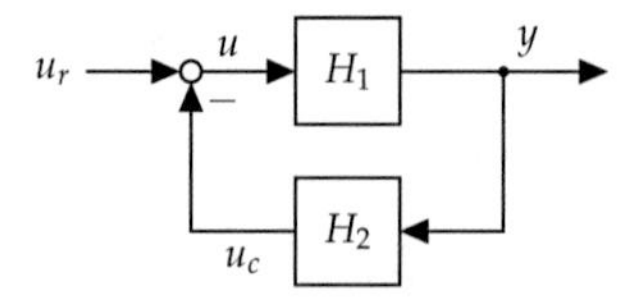

Figure A.1. Output in the input-/output framework. If the mapping H_1 is passive and the controller H_2 is strictly passive one concludes that $y \in \mathcal{L}$.

Izhikevich (1997) and references therein. The notion of reduced phase models greatly simplifies the system to be controlled. The main statement of the concept of reduced phase models is the following: Consider a family of dynamical systems of the form

$$\dot{\xi}(t) = \hat{f}(\xi(t)) \,, \quad \xi(t) \in \mathbb{R}^n \,, \tag{A.10}$$

with an exponentially stable limit cycle $\gamma \subset \mathbb{R}^n$ with period T. Then

$$\dot{x}(t) = \omega \,, \quad x(t) \in \mathcal{S}^1 \,, \tag{A.11}$$

is a local canonical model for such oscillators, where $x(t)$ is called the phase of the oscillator with frequency $\omega = \frac{2\pi}{T}$. This statement is based on the notion of *isochrons* introduced in A. T. Winfree (1974) and its basic idea, illustrated in Figure A.2, is to find a neighborhood W of γ and a function $\psi \colon W \to \mathcal{S}^1$, such that $x(t) = \psi(\xi(t))$ is a solution of Eq. (A.11). Winfree called the set of all initial conditions $z(0) \in \mathbb{R}^n$ of which the solution $z(t)$ approaches the solution $\xi(t)$, with $\xi(0) \in \gamma$, an *isochron* of $\xi(0)$ defined as

$$M_{\xi(0)} = \{z(0) \in W \colon \|\xi(t) - z(t)\| \to 0 \text{ as } t \to \infty\} \,. \tag{A.12}$$

Furthermore, Guckenheimer (1975) showed that there always exists a neighborhood W of a limit cycle that is invariantly foliated by the isochrons M_ξ, $\xi \in \gamma$, in the sense that the flow maps isochrons to isochrons. Consider the function $\psi_2 \colon W \to \gamma$, sending a point in the neighborhood $z \in M_\xi \subset W$ to the generator of its isochron $\xi \in \gamma$. Additionally, the periodic orbit of an oscillator is homeomorphic to the unit circle. One can therefore define the function $\psi_1 \colon \gamma \to \mathcal{S}^1$ which maps the solution $\xi(t)$ with $\xi(0) \in \gamma$ to the solution of Eq. (A.11). The function $\psi \colon W \to \mathcal{S}^1$ is a composition of ψ_1 and ψ_2, $\psi = \psi_1 \circ \psi_2$, mapping $\xi(t) \in W$ uniquely to its corresponding phase $x(t)$ of the reduced phase model (Figure A.2).

Applying the theory of reduced phase models to a weakly forced oscillator

$$\dot{\xi}(t) = f(\xi(t), u(t)) \,, \quad \xi(0) = \xi_0 \in W \,, \quad u(t) \in \mathbb{R} \tag{A.13}$$

where the term $u(t) = \varepsilon v(t)$ denotes an exogeneous input, one obtains the reduced phase model of the form

$$\dot{x}(t) = \omega + Z(x(t))\, u(t) \,. \tag{A.14}$$

Here, weakly forced refers to the situation where ε is sufficiently small such that $\xi(t)$ stays inside the neighborhood W for all $t > 0$. The function Z is called the *phase response curve* (PRC) and describes the magnitude of phase changes after perturbing an oscillatory system (Figure A.3). Based on *Malkin's Theorem* (Malkin 1949; Malkin 1956) the PRC is the solution of the adjoint problem $\mathrm{d}\, Z(t) / \mathrm{d}\, t = -\left(\mathrm{D} f(\xi(t))\right)^\top Z(t)$, with the normalization condition $Z(t) f(\xi(t)) = 1$ for any t, where $\mathrm{D}f$ is the Jacobian matrix which is evaluated along the. periodic orbit, $\xi(t) \in \gamma$.

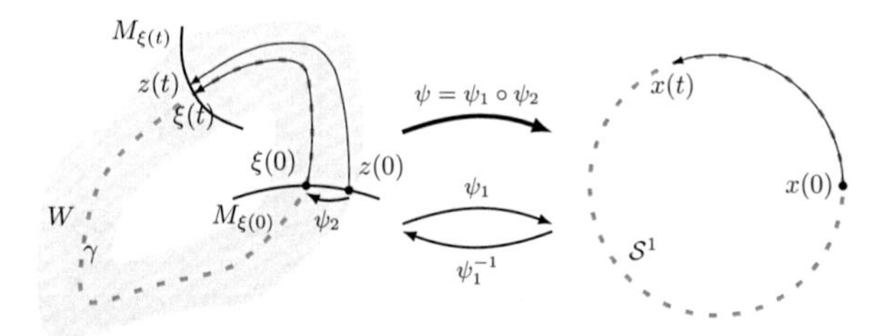

Figure A.2. A neighborhood W of the limit cycle γ of an oscillator is invariantly foliated by isochrons M_ξ. The flow maps isochrons to isochrons. The function $\psi = \psi_1 \circ \psi_2$ maps an oscillator $\xi(t) \in W$ uniquely to its phase on the unit circle $x(t) = \psi(\xi(t))$. © 2019 IEEE.

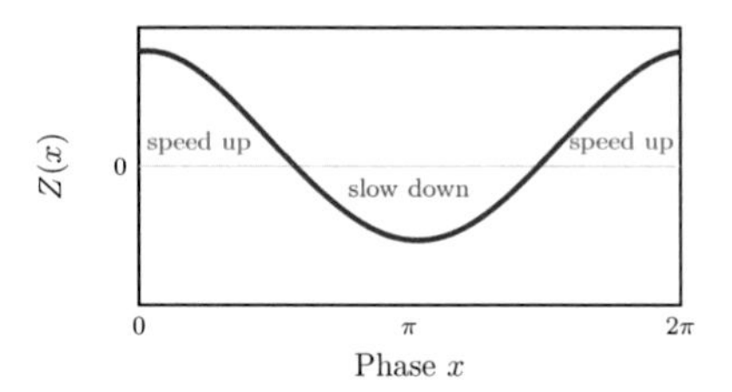

Figure A.3. Schematic phase response curve (PRC).

A.3 Fourier coefficients and circular moments

As will become apparent in this thesis, Fourier coefficients of the densities and of the phase response curve will play a crucial role in both the formulation and the solution of the ensemble control problem. This is not too surprising since all the functions involved in our problem setup can be viewed as 2π-periodic functions. We define the Fourier coefficients of a distribution by m_k:

$$p(\cdot,t) \sim m_k(p(\cdot,t)) := \frac{1}{2\pi}\int_0^{2\pi} p(x,t)\,\mathrm{e}^{-\mathrm{i}kx}\,\mathrm{d}x\,. \tag{A.15}$$

By omitting the argument in Eq. (A.15), we refer to the complex number $m_k = r\,\mathrm{e}^{\mathrm{i}k\phi}$, $r \in [0,1]$, $\phi \in \mathcal{S}^1$, obtained by evaluating $m_k(p(\cdot,t))$ with some specified distribution. Similarly, we have $q(t,\cdot) \sim \alpha_k$ for the desired distribution and $Z(\cdot) \sim c_k$ for the phase response curve. For better readability we will define a shorthand notation for frequently used expressions. Let $p(x,t)$ be a distribution function on the unit circle at time t. We will use $p_t(x) = p(x,t)$ especially when referring to the function in a L_2 inner product $\langle\cdot,\cdot\rangle$, i.e., we denote the L_2-norm of $p(x,t)$ by

$$\langle p_t, p_t\rangle = \int_0^{2\pi} p(x,t)p(x,t)\,\mathrm{d}x\,. \tag{A.16}$$

We will in addition use the term circular moments, when dealing with circular distributions, equally as Fourier coefficients as the circular moments are the 2π-scaled version of Fourier coefficients. In a synchronized population corresponding to a Dirac delta distribution, the length of the first circular moment $|m_1| = r$ is equal to one. This can be seen by using the sifting property of Dirac delta distributions

$$|m_1(\delta(\cdot-\mu))| = |\int_0^{2\pi} \delta(x-\mu)\,\mathrm{e}^{-\mathrm{i}x}\,\mathrm{d}x| = |1\,\mathrm{e}^{-\mathrm{i}\mu}| = 1\,. \tag{A.17}$$

Figure A.4 illustrates the relation between circular data, circular distributions and the first circular moment.

Remark A.1. Consider a function F on the unit circle $\mathcal{S}^1$. If F is 2π periodic, positive and absolutely continuous on $\mathcal{S}^1$ then the Fourier series of F converges uniformly and absolutely and

$$\langle F,F\rangle = \sum_{k=-\infty}^{\infty} |a_k| < \infty\,. \tag{A.18}$$

Proposition A.1. *For a continuous function $F \in C^\infty$ on the unit circle, there exists some $\nu_F < \infty$ such that $|a_k| = 0$ for all $k > \nu_F$.*

Definition A.2. *Let F be a continuous function with Fourier coefficients a_k for which Proposition A.1 holds. We define with ν_F the index of the largest non-zero Fourier coefficient of F*

$$\nu_F = \max\{k \in \mathbb{N} \mid a_k \neq 0\}\,. \tag{A.19}$$

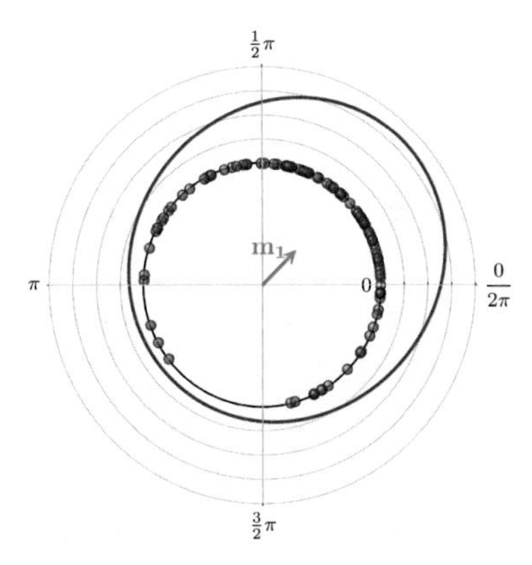

Figure A.4. Samples of circular data (blue), drawn from a von Mises distribution (red). The first circular moment is a complex number (green).

Appendix B

Experimental protocols and data preprocessing

B.1 Ordering cells in pseudotime

This section presents the description of cell biology with a focus on single cell data and the class of trajectory inference algorithms which are the starting point several results within this thesis.

Observing a snapshot of individual cells spread over different stages of a dynamical process can reveal what the process would look like for a single cell over time. Recovering a dynamical process form a snapshot of individual cells can be separated in two tasks: (1) ordering the cells, identifying a sequence of molecular events and (2) identifying the timing or the velocity of the trajectory. Trajectory infer (TI) were developed to order cells in snapshot data thereby solving the first task. In the following we review the basic concept of trajectory inference algorithms and pseudotime and refer the interested reader to comprehensive reviews and analyses in Saelens et al. (2019), Tritschler et al. (2019) and Luecken and Fabian J Theis (2019).

The development of powerful single cell experimental methods like CyTOF and single cell RNA-sequencing called for new computational methods to analyse the ever growing data sets (Figure B.1). As cellular diversity cannot sufficiently described by a discrete classification system such as clustering, TI algorithms were developed to capture the continuous nature of biological processes that generate the observed heterogeneity (Tanay and Regev 2017). Trajectory inference methods interpret single cell data as snapshot of a continuous process. The algorithms then reconstruct this process by finding paths through cellular space that minimize the difference between neighboring cells. The ordering of cells along the path is described in pseudotime (Figure B.2). The pseudotime value is a measure of how much progress an individual cell has made through a process such as differentiation or the cell cycle. Monocle (Trapnell et al. 2014) and Wanderlust (Bendall et al. 2014) established the TI field and the number of available methods grew rapidly in recent years. The different methods, comprehensively compared in Saelens et al. (2019), vary in their conceptual approach, their complexity and the used metric. Each method has its strengths and weaknesses and any inferred trajectory should be confirmed with an alternative method to revise the obtained cell order. Key to the success of any trajectory inference method are the input features, which must be informative for the developmental process we wish to model (Tritschler et al. 2019). While these methods often successfully find the correct sequence, they perform poorly when it comes to the actual dynamics as pseudotime is an abstract unit of progress, depending on various factors such as the distance metric and most prominently the labels from the measurements. A good way to test the consistency of the result from different pseudotime algorithms is by plotting the pseudotime values against each other (Figure B.3).

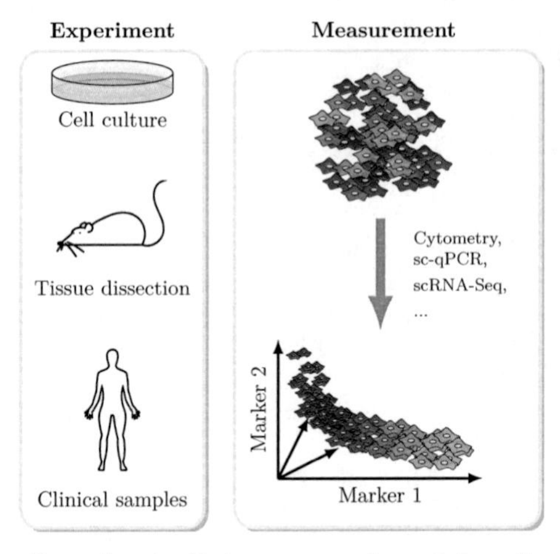

Figure B.1. Single-cell experiments of heterogeneous cell populations. Experimental samples are drawn from, cell culture, tissue dissection or biopsy. A biological process is causing heterogeneity in the cell population. Cells from single-cell experiments of a heterogeneous population are distributed around a lower dimensional process manifold in dataspace.

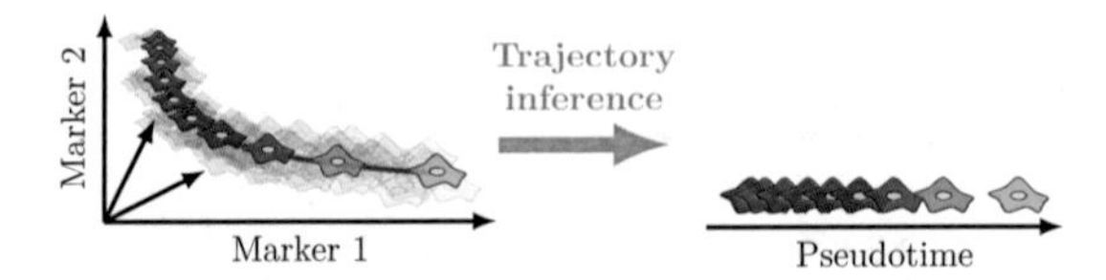

Figure B.2. Trajectory inference algorithms reconstruct biological processes by finding paths through cellular space. The ordering of cells along the path is described in pseudotime. Pseudotime value is a proxy for the progress of an individual cell in a continuous biological process.

The algorithms produced the same cell ordering of cells if the slope is positive. A diagonal line means that the pseudotime values are also identical. If different markers or pseudotime algorithms maintain the positive slope, then chances are high that the identified order is correct, despite possible deviations from the diagonal. MAPiT will in turn provide the right real-scale trajectory unaffected by the choice of algorithm or markers.

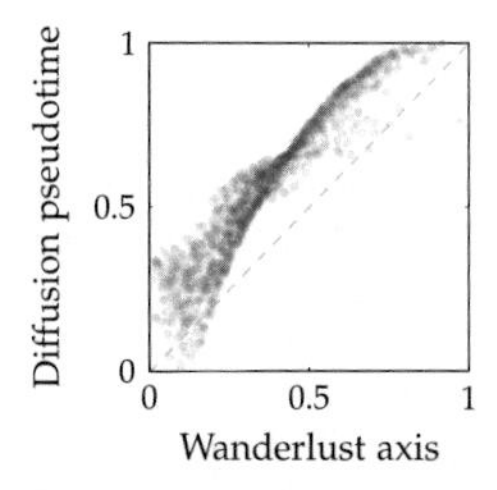

Figure B.3. Pseudotime values vary for different algorithms. In this example, the order of cells was largely conserved in both algorithms (monotonously increasing data points), however the pseudotime values were clearly different between both algorithms (deviation from the diagonal).

The results in the this thesis build upon the results of pseudotime algorithms. However, the developed methods and concepts are unaffected by the choice of the actual pseudotime algorithm used to generate a cellular order, as we show in Chapters 2 and 3.

B.2 Establishment of NCI-H460/geminin cell line

The human lung cancer cell line NCI-H460 was obtained from the American Type Culture Collection (LGC Standards GmbH, #HTB-177). Cells were maintained in RPMI 1640 medium supplemented with 5 % fetal calf serum at 37°C in a humidified incubator with 5 % CO_2. Fluorescent ubiquitination-based cell cycle indicator (Fucci) probe mAG-hGem(1/110) was obtained from RIKEN Brain Science Institute, Japan (Sakaue-Sawano et al. 2008) and subcloned into pEFpuro. Cells were transfected with the expression plasmid using Lipofectamine 2000 (Life Technologies) according to the manufacturers protocol. Stable transformants were selected using 0.3 $\frac{\mu g}{ml}$ puromycin and single cell selection was performed after two weeks. Several single cell clones were imaged for their total cell cycle length and length of G1 or S/G2/M phases by time lapse fluorescence microscopy using the Cell Observer system (Zeiss) equipped with a humidified imaging chamber at 37°C and 5 % CO_2. A single clone with cyclic expression of mAG-hGeminin and cell cycle kinetics similar to wild type cells was selected for subsequent experiments and referred to as NCI-H460/geminin.

B.3 Live cell microscopy

NCI-H460/geminin cells were imaged for their total cell cycle length and length of G1 or S/G2/M phases by time lapse fluorescence microscopy using the Cell Observer system (Zeiss) equipped with a humidified imaging chamber at 37° C and 5 % CO_2. Randomly chosen cells were manually tracked and the length of G1 phase and total cell cycle length extracted. The time of G1 to S phase transition was defined to be the first video frame with visible geminin signal in a cell.

B.4 Flow cytometric analysis

NCI-H460/geminin cells were harvested using trypsine/EDTA, fixed with 4 % PFA and permeabilized with FACS™permeabilization solution (BD Bioscience) for 30 min at room temperature. After washing with PBS/3 % FCS, 2×10^5 cells were resuspended in PBS/3 % FCS containing the primary antibody (Phospho-Akt (Ser473) (D9E) XP® Rabbit mAb (Cell Signalling Technology) or Cyclin B1 (D5C10) XP® Rabbit mAb (Cell Signalling Technology)). After incubation for 1 h at room temperature, cells were washed two times with PBS/3 % FCS, subsequently resuspended in PBS/3 % FCS containing the secondary antibody (Alexa Fluor 647-conjugated AffiniPure $F(ab')_2$ Fragment Goat Anti-Rabbit IgG (H+L) (Jackson ImmunoResearch)) and incubated for 45 min at room temperature. Cells were washed again, resuspended in PBS containing DNA counterstain DAPI (Applichem) and incubated for 10 min at room temperature. Afterwards, cells were washed, resuspended in PBS and analyzed on the MACSQuant analyser 10 (Miltenyi Biotec) equipped with three lasers (405 nm, 40 mW; 488 nm, 30 mW; 638 nm, 20 mW). The chromophore Alexa 647 was excited at 638 nm and the fluorescence was registered through a 655-730 nm band pass filter, mAG-hGeminin was excited at 488 nm and detected by a 525/50 band pass filter and DAPI was excited at 405 nm and registered through a 450/50 nm band pass filter.

B.5 Construction of the average cell cycle trajectory

The construction of the average cell cycle progression path in the DAPI/log(geminin) data space was performed as proposed in the supplementary materials of Kafri et al. (2013) with a recursive search algorithm. The start point, is chosen as the global density maximum. The trajectory through the dataset is then generated by iterative finding the point with the maximal cell density on the perimeter of a circle centered on the previously identified maximum. The generated trajectory is terminated once the DNA content of the maximum on the circle perimeter exceeds two times the DNA content of the start value. The generated trajectory can be seen as a function $l : \mathbf{y_c} \mapsto s$ which maps the cell cycle related readouts to a position on the path. The kernel density estimation was done by using a diffusion based approach which assumes a Gaussian kernel with optimal bandwidth selection (Botev et al. 2010). The association of single cell measurements to the curve was done by calculating the probability density for a single data point to be on a position s of the average cell cycle trajectory. To this end for each cell a 2D Gaussian centered on the data coordinates with the bandwidth obtained from the kernel density estimation $\rho^i_{y_c}(\mathbf{y_c})$ is evaluated on the constructed trajectory and normalized to one to generate a probability density

$$\rho^i_s(s) = \rho^i_{y_c}\left(l^{-1}(s)\right)\left(\int_0^1 \rho^i_{y_c}\left(f^{-1}(\sigma)\right)\,\mathrm{d}\sigma\right)^{-1}. \tag{B.1}$$

The cell number density along the path is then given by the sum of all individual probability densities normalized by the number of data points/cells

$$\rho_s(s) = \frac{1}{N}\sum_{i=1}^{N}\rho_s^i(s). \tag{B.2}$$

B.6 Estimation of constant speed and noise from cell cycle lengths

In the simplest case speed and noise functions of the SDE in x are just constants v and D which can be identified from measurements of total cell cycle length of a population of cells. The solution of the corresponding PDE Eq. (3.28) with reflecting and absorbing boundary at $x = 0$ and $x = 1$ provides when initialized with a Dirac delta distribution at the origin $n_0(x) = \delta(x)$ a probability measure for a cell to have age t given position x on the path. The probability distribution of the total cell cycle lengths $p(T)$ is the first passage time distribution of cells over the boundary $x = 1$, which is by definition the end of the cell cycle. The first passage time equals the flux at across the boundary (Redner 2001).

$$J(t, x = 1) \to p(T) \tag{B.3}$$

The solution is obtained by constructing an approximate analytic form of Green's function of the PDE Eq. (3.28) with absorbing and reflecting boundary. Green's function $G_r(x_0; x, t)$ for a FPE with reflecting boundary at $x = 0$ is constructed by a sum of a principal solution $U_r(x_0; x, t)$ (representing the solution in an infinite domain) and a correcting regular solution $g_r(x_0; x, t)$ (Singer et al. 2008).

$$G_r(x_0; x, t) = U_r(x_0; x, t) + g_r(x_0; x, t). \tag{B.4}$$

The principal solution is the standard solution for biased diffusion in an unbounded domain

$$U_r(x_0; x, t) = \frac{1}{\sqrt{2\pi D^2 t}} \exp\left(-\frac{(x - x_0 - vt)^2}{2D^2 t}\right). \tag{B.5}$$

By the inspection approach the regular solution is constructed as an mirrored image of the principal solution with correction term due to the drift (Green 1988).

$$\begin{aligned} g_r(x_0; x, t) = & \frac{1}{\sqrt{2\pi D^2 t}} \left(\exp\left(-\frac{(x + x_0 - vt)^2}{2D^2 t}\right) \exp\left(-\frac{2vx_0}{D^2}\right)\right) \\ & - \frac{v}{D^2} \exp\left(\frac{2vx}{D^2}\right) \operatorname{erfc}\left(\frac{x + x_0 + vt}{\sqrt{2D^2 t}}\right). \end{aligned} \tag{B.6}$$

To construct Green's function with the additional absorbing boundary at $x = 1$, $G(x_0; x, t)$, we again use the inspection approach and subtract an adjusted mirrored image of the principal solution for which we take Green's function with reflecting boundary at $x = 0$ derived above.

$$G(x_0; x, t) = G_r(x_0; x, t) - \gamma(x, t)\, G_r(x_0; -x + 2, t). \tag{B.7}$$

The correcting factor $\gamma(x, t)$ is chosen such that the absorbing boundary condition is fulfilled.

$$\gamma(x, t) = \frac{G_r(1; x, t)}{G_r(1; -x + 2, t)} \tag{B.8}$$

The approximation of Green's function with one reflecting and one absorbing boundary as proposed is especially with positive drift and a small diffusion coefficient in good agreement with the true solution. The solution to the PDE Eq. (3.28) is obtained by direct use of the property of the Dirac delta function at $x = 0$ as initial density.

$$\begin{aligned} n(t,x) &= \int_0^1 G(\xi; x, t)\, n_0(\xi)\, \mathrm{d}\xi \\ &= \int_0^1 G(\xi; x, t)\, \delta(\xi)\, \mathrm{d}\xi \\ &= G(0; x, t) \end{aligned}$$

Differentiating with respect to x and evaluation of the flux at $x = 1$ provides the expression for the distribution of the cell cycle length

$$p(T) = -J(T,x)|_{x=1} = -\frac{1}{2}D^2 \left. \frac{\partial n(t,x)}{\partial x} \right|_{x=1} . \tag{B.9}$$

The optimization problem for the identification of v and D, given a set of total cell cycle length measurements $\mathcal{D} = \{T^1\}_{i=1,\dots,n}$ is then to maximize the probability to observe $\mathcal{D}$ with the model Eq. (3.28).

$$\underset{v,D}{\operatorname{argmax}} \prod_{i=1}^{n} p(T^i) . \tag{B.10}$$

Appendix C

Spheroid growth model

In order to correctly calculate the transformation from pseudotime to the distance scale the cumulative density of cells in the spheroid w.r.t. the distance from the surface must be known. This requires (1) the size of the spheroid and (2) the size at which the necrotic core emerges (Figure C.1). Both quantities can be obtained from models of spheroid growth dynamics.

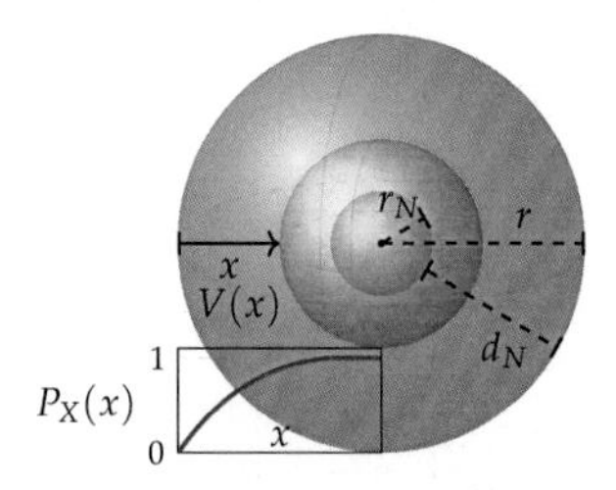

Figure C.1. Sphere geometry. Volume of spherical shell $V(x)$ with radius r and thickness x (blue), normalised to the volume of the sphere $V_S(r)$ minus the volume of the necrotic core $V_S(r - r_N)$ provides the cumulative distribution of cells in the spheroid $P_X(x)$.

The spheroid volume as a function of time can be calculated with a spheroid growth model with solely one layer of proliferative cells at the spheroid surface (Conger and Ziskin 1983). However, emergence of a necrotic core is not captured with such a simple model but has to be estimated from experimental data. Spheroids are seen as analogous to avascular tissue or tumour mass and diffusion limitation to many molecules, particularly O_2, becomes apparent in spheroids with diameters larger $150 - 200\,\mu m$ (Lin and Chang 2008). In addition, inefficient mass transport leads to metabolic waste accumulation inside the spheroids. Therefore, spheroids with a diameter larger $500\,\mu m$ commonly display a layer-like structure comprising a necrotic core surrounded by a viable rim. The viable rim consists of an inner layer of quiescent cells and an outer layer of proliferating cells (Lin and Chang 2008; D. V. LaBarbera et al. 2012). We developed a layer-based growth model to identify the spheroid size at which the necrotic core emerges. The model consists of $i \in \{1, \ldots, I\}$ different layers of which each layer has its own characteristics determined by the parameters of the layer:

- d_i layer thickness in μm
- γ_i fraction of proliferating cells in layer i
- μ_i growth rate in $\frac{1}{\text{day}}$ of the proliferating cells in layer i

Furthermore we have two parameters that are independent of the layers:

- r_0 initial spheroid radius in μm.
- v_c volume of a single cell in μm^3.

We defined the innermost layer as a necrotic core without any cells. The necrotic core then emerges at a spheroid radius which is equal to the sum of the estimated thickness of the outer layers

$$r_N = \sum_{i=1}^{I-1} d_i \,. \tag{C.1}$$

Parameters d_i, γ_i, μ_i thus are sufficient to describe the growth of a spheroid. The volume growth of a spheroid is given as the sum of the volume change in each layer by:

$$\begin{aligned} \frac{\mathrm{d}V}{\mathrm{d}t} &= \sum_{i=1}^{I} V_i \gamma_i \mu_i \\ &= \frac{4}{3}\pi \sum_{i=1}^{I} \gamma_i \mu_i \left((r - d_{i-1})^3 - (r - d_{i-1} - d_i)^3 \right) \,. \end{aligned} \tag{C.2}$$

The change of the volume can furthermore be decomposed with the chain rule into

$$\frac{\mathrm{d}V}{\mathrm{d}t} = \frac{\mathrm{d}V}{\mathrm{d}r}\frac{\mathrm{d}r}{\mathrm{d}t} \,. \tag{C.3}$$

We know that the change of the volume with respect to the radius is given by

$$\frac{\mathrm{d}V}{\mathrm{d}r} = 4\pi r^2 \,. \tag{C.4}$$

By inserting Eqs. (C.2) and (C.4) into Eq. (C.3) one solves for a change of the radius of the spheroid over time, which is sufficient to describe the whole spheroid growth

$$\frac{\mathrm{d}r}{\mathrm{d}t} = \frac{1}{3r^2} \sum_{i=1}^{I} \gamma_i \mu_i \left((r - d_{i-1})^3 - (r - d_{i-1} - d_i)^3 \right) \,. \tag{C.5}$$

This nonlinear ODE Eq. (C.5) can be solved numerically to obtain the spheroid radius over time $r(t)$. Other quantities of interest can then be computed:

- Spheroid volume: $V(t) = \frac{4}{3}\pi r(t)^3$
- Total cell number in the spheroid: $N(t) = \frac{V(t)}{v_c}$
- Volume of the single layers: $V_i(t) = \frac{4}{3}\pi \left(\left(r(t) - \sum_{j=1}^{i-1} d_j \right)^3 - \left(r(t) - \sum_{j=1}^{i} d_j \right)^3 \right)$
- Cell number per layer: $N_i(t) = \frac{V_i(t)}{v_c}$
- Number of proliferating cells in a layer: $N_i^P(t) = N_i(t)\gamma_i$
- Number of quiescent cells in a layer: $N_i^Q(t) = N_i(t)(1 - \gamma_i)$

Parameter estimation for the spheroid growth model

The parameters of the spheroid growth model were identified by comparing the model predictions to spheroid volumes obtained by microscopy and cell numbers obtained from flow cytometry counts of individual spheroids at different days post seeding. We used a weighted sum of squares as a measure for the difference between the dataset and the model predictions. The parameters were estimated initially using the CMA-ES algorithm (Hansen 2006). In a next step the parameters were sampled using a MCMC DRAM algorithm to additionally obtain the uncertainties of the parameter values (Haario et al. 2006). The models were initiated with different numbers of layers ranging from $I \in \{2,3,4\}$. Based on the Bayesian Information Criterion (BIC), the model with two layers followed by a necrotic core ($I = 3$) was identified as the most plausible. The spheroids then consisted of an outer proliferating layer with thickness of $20 - 25\,\mu m$, which is close to the estimated diameter of a single cell. The second layer has a thickness of roughly $250\,\mu m$ and does not contain proliferating cells. The necrotic core emerges at a radius of $r_N \approx 270$ which is in good agreement with literature values (Lin and Chang 2008; D. V. LaBarbera et al. 2012). Based on the estimated parameters a reduced ODE model for the spheroid radius is given by:

$$\frac{\mathrm{d}r}{\mathrm{d}t} = \mu_1 \left(d_1 - \frac{d_1^2}{r} - \frac{d_1^3}{r^2} \right) . \tag{C.6}$$

Once the spheroid reaches a certain size where $r >> d_1$ Eq. (C.6) reduces to a constant growth of the spheroid

$$\frac{\mathrm{d}r}{\mathrm{d}t} = \mu_1 d_1 , \tag{C.7}$$

which reflects the simple linear growth model of the spheroid radius already presented in Conger and Ziskin (1983). Our layer-based model thus could reproduce experimental data and is furthermore in line with previous spheroid growth models. The derived values for spheroid growth $r(t)$ and the radius at which the necrotic core emerges r_N were used to calculate the position-dependent cell density,

$$p_X(x) = 3\frac{(r-x)^2}{r^3 - r_N^3} , \tag{C.8}$$

required for the application of MAPiT to single-cell data of dissociated spheroids.

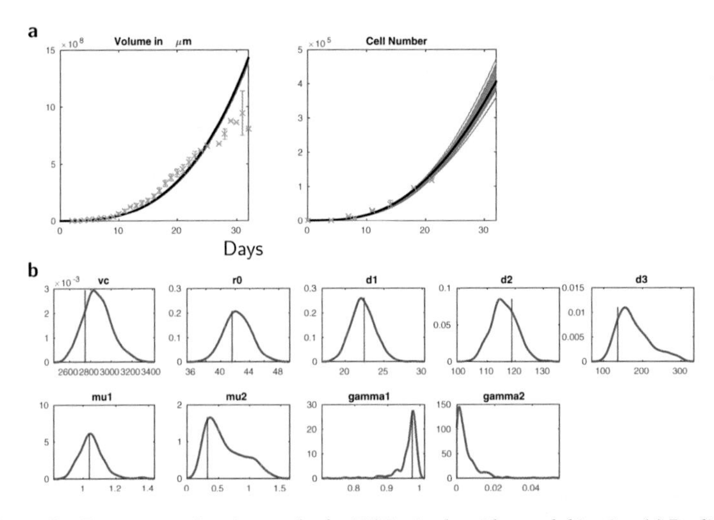

Figure C.2. Parameter estimation results for HCT116 spheroid growth kinetics. (**a**) Predictions of a three layer model with a necrotic core for spheroid volume and total cell number. Best fit (black) and 200 best samples (red) in comparison with experimental data (green). Model predictions correspond very well with the measurements. (**b**) Posterior parameter distributions and best fit (red) obtained by MCMC of a three layer spheroid growth model. Parameter *vc* and *ro* represent single cell volume and spheroid radius at $t = 0$. Parameters *d1*-*d3* define thickness of layers with distinct properties such as growth rates *mu1*, *mu2* and fraction of proliferating cells *gamma1*, *gamma2* in the corresponding layer. All parameters were identifiable.

Appendix D

Cell cycle model

Five-variable skeleton model of the mammalian cell cycle by C. Gérard and Goldbeter (2011) and Claude Gérard et al. (2012) consisting of the four main cyclin/Cdk complexes, the transcription factor E2F and the protein Cdc20 (Figure D.1). In the model, growth factors ensure the synthesis of the cyclin D/Cdk4–6 complex which rapidly reaches a steady state and promotes progression in the G1 phase by activating the transcription factor E2F. E2F causes activation of cyclin E/Cdk2 at the G1/S transition, and cyclin A/Cdk2 during S phase. During G2, cyclin A/Cdk2 triggers the activation of cyclin B/Cdk1, which leads to the G2/M transition. During mitosis, cyclin B/Cdk1 activates the protein Cdc20. Cdc20 creates a negative feedback loop involving cyclin A/Cdk2 and cyclin B/Cdk1 by promoting the degradation of these complexes. This negative feedback loop together with the positive loop of cyclin E/Cdk2 via E2F results in stable oscillations (Figure D.1 b).

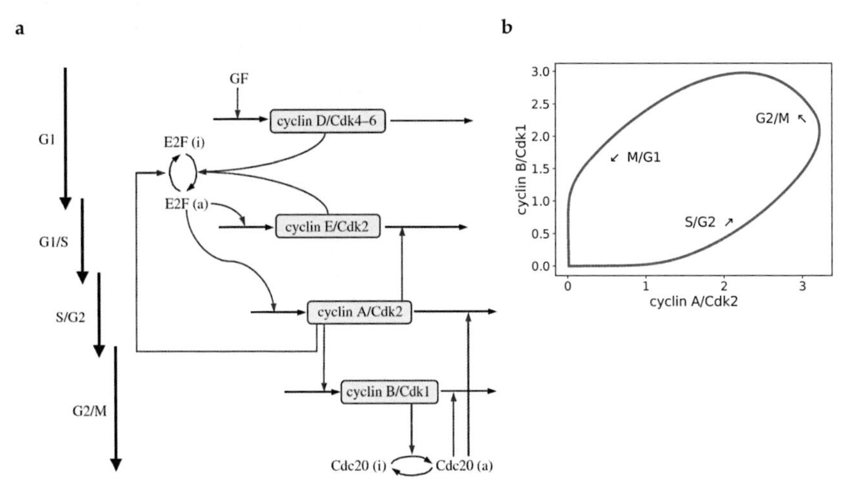

Figure D.1. Cell cycle model from C. Gérard and Goldbeter (2011). (**a**) Scheme of the Cdk network driving the mammalian cell cycle. (**b**) Limit cycle oscillations shown as a projection onto the cyclin A/Cdk2 versus cyclin B/Cdk1 phase plane.

The system of ODEs, with complexes cyclin A/Cdk2, cyclin B/Cdk1 and cyclin E/Cdk2 abbreviated by Ma, Mb and Me, respectively.

$$
\begin{aligned}
\frac{\mathrm{d}\,Ma}{\mathrm{d}t} &= v_{sa} \cdot E2F - V_{da} \cdot Cdc20 \cdot \frac{Ma}{K_{da} + Ma} \\
\frac{\mathrm{d}\,Mb}{\mathrm{d}t} &= v_{sb} \cdot Ma - V_{db} \cdot Cdc20 \cdot \frac{Mb}{K_{db} + Mb} \\
\frac{\mathrm{d}\,Me}{\mathrm{d}t} &= v_{se} \cdot E2F - V_{de} \cdot Ma \cdot \frac{Me}{K_{de} + Me} \\
\frac{\mathrm{d}\,E2F}{\mathrm{d}t} &= V_{1e2f} \cdot \frac{E2F_{tot} - E2F}{K_{1e2f} + (E2F_{tot} - E2F)} \\
\frac{\mathrm{d}\,Cdc20}{\mathrm{d}t} &= V_{1dc20} \cdot Mb \cdot \frac{Cdc20_{tot} - Cdc20}{K_{1cdc20} + (Cdc20_{tot} - Cdc20)} - V_{2cdc20} \cdot \frac{Cdc20}{K_{2cdc20} + Cdc20}
\end{aligned}
\tag{D.1}
$$

Appendix E

Technical computations and proofs

E.1 Speed equation

The function $\tau : s \mapsto a$ establishes a direct relation between the path position s and the age of a cell. To obtain it, one has to know the speed $f(s)$ in which the cells proceed on the cell cycle path. Without diffusion, the PDE Eq. (3.8) reduces to

$$\frac{\partial n(t,s)}{\partial t} = -\frac{\partial}{\partial s} J(t,s) = -\frac{\partial}{\partial s}(f(s)n(t,s)).$$

In a stationary case, where $n(t,s) = N(t)p_s(s)$, we can derive an equation for $f(s)$ in a few steps:

1. Integrate from 0 to s:

$$\frac{\partial}{\partial t} \int_0^s n(t,\sigma)\,\mathrm{d}\sigma = J(t,s) + k$$

2. The left-hand side is equal to the population growth in Eq. (3.14):

$$\gamma \int_0^s n(t,\sigma)\,\mathrm{d}\sigma = J(t,s) + k = -f(s)n(t,s) + k$$

3. Solve for the flux $J(t,s)$:

$$J(t,s) = \gamma \int_0^s n(t,\sigma)\,\mathrm{d}\sigma - k$$

4. Solve for the $f(s)$:

$$f(s) = \frac{k - \gamma \int_0^s n(t,\sigma)\,\mathrm{d}\sigma}{n(t,s)}$$

5. Expand fraction with $N(t)^{-1} = \left(\int_0^1 n(t,\sigma)\,\mathrm{d}\sigma \right)^{-1}$:

$$f(s) = \frac{\frac{k}{N(t)} - \gamma\, P(s)}{\rho_s(s)}$$

6. Identify k by employing the boundary condition for a dividing population (Eq. (3.12)):

$$J(t,0) = 2J(t,1)$$
$$\gamma \int_0^0 n(t,\sigma)\,\mathrm{d}\sigma - k = 2\gamma \int_0^1 n(t,\sigma)\,\mathrm{d}\sigma - 2k$$
$$-k = 2\gamma N(t) - 2k$$
$$k = 2\gamma N(t)$$

Finally, by substituting k we get for the cell speed along the path the following equation:

$$f(s) = \gamma \frac{2 - P(s)}{\rho_s(s)} \tag{E.1}$$

E.2 Derivation of ODE for $\varphi : s \mapsto x$

The noise factor $g(s)$ for the s-parameterisation can be calculated with a function $\varphi : s \mapsto x$ which transforms the SDEs Eqs. (3.7) and (3.27). From the Itô differentiation rule we get for such a function the following SDE:

$$\mathrm{d}\varphi = \left(\frac{\partial \varphi}{\partial s} f(s) + \frac{1}{2} g^2(s) \frac{\partial^2 \varphi}{\partial s^2} \right) \mathrm{d}t + \frac{\partial \varphi}{\partial s} g(s)\,\mathrm{d}W \tag{E.2}$$

From a coefficient comparison of Eqs. (3.27) and (E.2) we get:

$$\frac{\partial \varphi}{\partial s} g(s) = h(\varphi) \qquad \rightarrow \qquad g(s) = \frac{h(\varphi)}{\frac{\partial \varphi}{\partial s}} \tag{E.3}$$

$$\frac{\partial \varphi}{\partial s} f(s) + \frac{1}{2} g(s)^2 \frac{\partial^2 \varphi}{\partial s^2} = v(\varphi) \tag{E.4}$$

We can now substitute the noise in Eq. (3.26) with Eq. (E.3) and evaluate the derivatives in the equation

$$\begin{aligned}
f(s) &= \gamma \frac{2 - P(s)}{\rho_s(s)} + \frac{1}{2\rho_s(s)} \left(\frac{h(\varphi)}{\frac{\partial \varphi}{\partial s}} \right)^2 \frac{\partial \rho_s(s)}{\partial s} + \frac{h(\varphi)}{\frac{\partial \varphi}{\partial s}} \frac{\partial}{\partial s} \left(\frac{h(\varphi)}{\frac{\partial \varphi}{\partial s}} \right) \\
&= \gamma \frac{2 - P(s)}{\rho_s(s)} + \frac{h^2(\varphi)}{2\rho_s(s) \left(\frac{\partial \varphi}{\partial s} \right)^2} \frac{\partial \rho_s(s)}{\partial s} + \frac{h(\varphi)}{\frac{\partial \varphi}{\partial s}} \frac{\frac{\partial \varphi}{\partial s} \frac{\partial h(\varphi)}{\partial s} - h(\varphi) \frac{\partial^2 \varphi}{\partial s^2}}{\left(\frac{\partial \varphi}{\partial s} \right)} \\
&= \gamma \frac{2 - P(s)}{\rho_s(s)} + \frac{h^2(\varphi)}{2\rho_s(s) \left(\frac{\partial \varphi}{\partial s} \right)^2} \frac{\partial \rho_s(s)}{\partial s} + \frac{h(\varphi)}{\frac{\partial \varphi}{\partial s}} \frac{\left(\frac{\partial \varphi}{\partial s} \right) \frac{\partial h(\varphi)}{\partial \varphi} - h(\varphi) \frac{\partial^2 \varphi}{\partial s^2}}{\left(\frac{\partial \varphi}{\partial s} \right)} \\
&= \gamma \frac{2 - P(s)}{\rho_s(s)} + \frac{h^2(\varphi)}{2\rho_s(s) \left(\frac{\partial \varphi}{\partial s} \right)^2} \frac{\partial \rho_s(s)}{\partial s} + \frac{h(\varphi)}{\frac{\partial \varphi}{\partial s}} \frac{\partial h(\varphi)}{\partial \varphi} - \frac{h^2(\varphi)}{\left(\frac{\partial \varphi}{\partial x} \right)^3} \\
f(s) &= \gamma \frac{2 - P(s)}{\rho_s(s)} + \left(\frac{\partial \varphi}{\partial s} \right)^{-1} h(\varphi) \frac{\partial h(\varphi)}{\partial \varphi} + \left(\frac{\partial \varphi}{\partial s} \right)^{-2} \frac{h^2(\varphi)}{2\rho_s(s)} \frac{\partial \rho_s(s)}{\partial s} - \left(\frac{\partial \varphi}{\partial x} \right)^{-3} h^2(\varphi) \frac{\partial^2 \varphi}{\partial s^2}
\end{aligned} \tag{E.5}$$

In the next step, one eliminates $g(s)$ and $f(s)$ in Eq. (E.4) with the results Eq. (E.3) and Eq. (E.5).

$$\frac{\partial\varphi}{\partial s}\left(\gamma\frac{2-P(s)}{\rho_s(s)}+\left(\frac{\partial\varphi}{\partial s}\right)^{-1}h(\varphi)\frac{\partial h(\varphi)}{\partial\varphi}+\left(\frac{\partial\varphi}{\partial s}\right)^{-2}\frac{h^2(\varphi)}{2\rho_s(s)}\frac{\partial\rho_s(s)}{\partial s}-\left(\frac{\partial\varphi}{\partial s}\right)^{-3}h^2(\varphi)\frac{\partial^2\varphi}{\partial s^2}\right)$$
$$+\left(\frac{\partial\varphi}{\partial s}\right)^{-2}\frac{h^2(\varphi)}{2}\frac{\partial^2\varphi}{\partial s^2}=v(\varphi)$$

For a better presentation we now define $\dot{\varphi}:=\frac{\partial\varphi}{\partial s}$ and $\dot{\varphi}:=\frac{\partial^2\varphi}{\partial s^2}$

$$v(\varphi)=\dot{\varphi}\left(\gamma\frac{2-P(s)}{\rho_s(s)}+(\dot{\varphi})^{-1}h(\varphi)\frac{\partial h(\varphi)}{\partial\varphi}+(\dot{\varphi})^{-2}\frac{h^2(\varphi)}{2\rho_s(s)}\frac{\partial\rho_s(s)}{\partial s}-(\dot{\varphi})^{-3}h^2(\varphi)\dot{\varphi}\right)$$
$$+(\dot{\varphi})^{-2}\frac{h^2(\varphi)}{2}\dot{\varphi}$$

and solve the equation for $\dot{\varphi}$

$$\dot{\varphi}=\dot{\varphi}\frac{1}{\rho_s(s)}\frac{\partial\rho_s(s)}{\partial s}+\dot{\varphi}^2\,2\left(\frac{1}{h(\varphi)}\frac{\partial h(\varphi)}{\partial\varphi}-\frac{v(\varphi)}{h^2(\varphi)}\right)+\dot{\varphi}^3\,\frac{2\gamma}{h^2(\varphi)}\frac{2-P(s)}{\rho_s(s)} \tag{E.6}$$

This second order nonlinear ODE for $\varphi(s)$ can be solved numerically as a boundary value problem. From the definition of s and x with equality at the start and the end, one gets the boundary as $\varphi(0)=0$ and $\varphi(1)=1$.

For the case with constant speed v and noise D in x parameterisation, the ODE Eq. (E.6) simplifies to

$$\dot{\varphi}=\dot{\varphi}\frac{1}{\rho_s(s)}\frac{\partial\rho_s(s)}{\partial s}-\dot{\varphi}^2\,2\frac{v}{D^2}+\dot{\varphi}^3\,\frac{2\gamma}{D^2}\frac{2-P(s)}{\rho_s(s)} \tag{E.7}$$

Having solved the ODE for $\varphi(s)$ one can calculated $g(s)$ from Eq. (E.3) and subsequently $f(s)$ from Eq. (3.26). By doing this all functions which are needed for the transformation of $\rho_s(s)$ to $\rho_a(a)$ are identified.

E.3 Calculation of division rate from cell cycle length distribution

For the identification of the age-dependent division rate $\alpha(a)$ in the steady state distribution of age-structured population models (Eq. (3.5)) one can exploit the fact, that the product of the steady state age-structure and the division rate must be equal to the cell cycle length distribution of a cell population.

$$p_c(a)=\rho_a(a)\alpha(a) \tag{E.8}$$

Substituting $\rho_a(a)$ with Eq. (3.5) we get the following expression

$$p_c(a)=2\gamma e^{-\gamma a}e^{-\int_0^a\alpha(\tilde{a})\,\mathrm{d}\tilde{a}}\alpha(a) \tag{E.9}$$

Taking the logarithm and differentiating with respect to a results in nonlinear ordinary differential equation for the logarithm of the division rate

$$\frac{\mathrm{d}}{\mathrm{d}a}\ln(\alpha(a))=\gamma+\alpha(a)+\frac{\mathrm{d}}{\mathrm{d}a}\ln(p_c(a)) \tag{E.10}$$

which can be solved numerically as γ and $p_c(a)$ are known from the experiments.

E.4 Proof of Lemma 6.1

The starting point is to consider

$$\frac{\mathrm{d}}{\mathrm{d}t}V(t) = \frac{1}{2}\int_0^{2\pi} \partial_t(\Delta_t^2)\,\mathrm{d}x = \int_0^{2\pi} \Delta_t\,(\partial_t\Delta_t)\,\mathrm{d}x. \tag{E.11}$$

With $\Delta_t = p_t - q_t$, as well as Eqs. (6.3) and (6.5), we have

$$\begin{aligned}(\partial_t\Delta_t)(x) &= -\partial_x\left((\omega + Z(x)u(t))p_t(x)\right) + \partial_x\left(\omega q_t(x)\right)\\ &= -\omega(\partial_x\Delta_t) - \partial_x(Z(x)p_t(x))u(t).\end{aligned}$$

Plugging the result in Eq. (E.11), we obtain

$$\frac{\mathrm{d}}{\mathrm{d}t}V(t) = -\omega\int_0^{2\pi} \Delta_t(\partial_x\Delta_t)\,\mathrm{d}x - \left(\int_0^{2\pi} \Delta_t\partial_x(Zp_t)\,\mathrm{d}x\right)u(t)$$

Here the first term can be seen to be equal to $(\omega/2)\int_0^{2\pi}\partial_x(\Delta_t)^2\,\mathrm{d}x = (\omega/2)[\Delta_t^2]_{t=0}^{t=2\pi} = 0$, where the vanishing is due to the 2π-periodicity of Δ_t^2. The final result is obtained via an integration by parts for the second term.

$$\frac{\mathrm{d}}{\mathrm{d}t}V(t) = \underbrace{-\omega\int_0^{2\pi} \Delta_t(\partial_x\Delta_t)\,\mathrm{d}x}_{=0} - \left(\int_0^{2\pi} \Delta_t\partial_x(Zp_t)\,\mathrm{d}x\right)u(t) \tag{E.12}$$

$$= -\underbrace{[\Delta_t Z_t p_t]_0^{2\pi}}_{=0}\,u(t) + \left(\int_0^{2\pi} \partial_x\Delta_t Z p_t\,\mathrm{d}x\right)u(t) \tag{E.13}$$

$$= \left(\int_0^{2\pi} (\partial_x\Delta_t) Z p_t\,\mathrm{d}x\right)u(t)\,. \tag{E.14}$$

Bibliography

Achim, Kaia, Jean-Baptiste Pettit, Luis R. Saraiva, Daria Gavriouchkina, Tomas Larsson, Detlev Arendt and John C. Marioni (2015). 'High-throughput spatial mapping of single-cell RNA-seq data to tissue of origin'. In: *Nature Biotechnology* 33.5, pp. 503–509. DOI: 10.1038/nbt.3209.

Arnol, Damien, Denis Schapiro, Bernd Bodenmiller, Julio Saez-Rodriguez and Oliver Stegle (2019). 'Modeling Cell-Cell Interactions from Spatial Molecular Data with Spatial Variance Component Analysis'. In: *Cell Reports* 29.1, 202–211.e6. DOI: 10.1016/j.celrep.2019.08.077.

Ashwin, Peter, Stephen Coombes and Rachel Nicks (2016). 'Mathematical Frameworks for Oscillatory Network Dynamics in Neuroscience'. In: *The Journal of Mathematical Neuroscience* 6.1, p. 2. DOI: 10.1186/s13408-015-0033-6.

Bandura, Dmitry R. et al. (2009). 'Mass Cytometry: Technique for Real Time Single Cell Multitarget Immunoassay Based on Inductively Coupled Plasma Time-of-Flight Mass Spectrometry'. In: *Analytical Chemistry* 81.16. PMID: 19601617, pp. 6813–6822. DOI: 10.1021/ac901049w.

Banfalvi, G. (2011). 'Overview of Cell Synchronization'. In: *Cell Cycle Synchronization: Methods and Protocols*. Ed. by Gaspar Banfalvi. Totowa, NJ: Humana Press, pp. 1–23.

Bendall, Sean C et al. (2014). 'Single-cell trajectory detection uncovers progression and regulatory coordination in human B cell development.' In: *Cell* 157.3, pp. 714–25. DOI: 10.1016/j.cell.2014.04.005.

Bergen, Volker, Marius Lange, Stefan Peidli, F. Alexander Wolf and Fabian J. Theis (2019). 'Generalizing RNA velocity to transient cell states through dynamical modeling'. In: *bioRxiv*. DOI: 10.1101/820936.

Billy, Frédérique, Jean Clairambault, Franck Delaunay, Céline Feillet and Natalia Robert (2013). 'Age-structured cell population model to study the influence of growth factors on cell cycle dynamics'. In: *Mathematical Biosciences and Engineering* 10.1, pp. 1–17.

Billy, Frédérique, Jean Clairambault, Olivier Fercoq, Stéphane Gaubert, Thomas Lepoutre and Thomas Ouillon (2011). 'Proliferation in Cell Population Models with Age Structure'. In: *AIP Conference Proceedings* 1389.1, pp. 1212–1215.

Botev, Z. I., J. F. Grotowski and D. P. Kroese (2010). 'Kernel density estimation via diffusion'. In: *Ann. Statist.* 38.5, pp. 2916–2957.

Brockett, Roger (1976). 'Nonlinear systems and differential geometry'. In: *Proceedings of the IEEE* 64.1, pp. 61–72. DOI: 10.1109/PROC.1976.10067.

Brockett, Roger (2010). 'On the control of a flock by a leader'. In: *Proceedings of the Steklov Institute of Mathematics* 268.1, pp. 49–57. DOI: 10.1134/S0081543810010050.

Brockett, Roger (2012). 'Notes on the Control of the Liouville Equation'. In: *Lect. Notes Math.* Vol. 2048. Lecture Notes in Mathematics. Springer Berlin Heidelberg, pp. 101–129.

Carmona-Alcocer, Vania, John H. Abel, Tao C. Sun, Linda R. Petzold, Francis J. Doyle, Carrie L. Simms and Erik D. Herzog (2018). 'Ontogeny of Circadian Rhythms and Synchrony in the Suprachiasmatic Nucleus'. In: *The Journal of Neuroscience* 38.6, pp. 1326–1334. DOI: 10.1523/JNEUROSCI.2006-17.2017.

Clairambault, J., S. Gaubert and Th. Lepoutre (2009). 'Comparison of Perron and Floquet Eigenvalues in Age Structured Cell Division Cycle Models'. In: *Mathematical Modelling of Natural Phenomena* 4.3, pp. 183–209.

Clairambault, Jean, Stéphane Gaubert and Thomas Lepoutre (2011). 'Circadian rhythm and cell population growth'. In: *Mathematical and Computer Modelling* 53.7-8, pp. 1558–1567.

Clairambault, Jean, Philippe Michel and Benoît Perthame (2007). 'A Mathematical Model of the Cell Cycle and Its Circadian Control'. In: *Math. Model. Biol. Syst.* Birkhäuser Boston, pp. 239–251.

Conger, A D and Marvin C. Ziskin (1983). 'Growth of mammalian multicellular tumor spheroids.' In: *Cancer Research* 43.2, pp. 556–60.

Csikasz-Nagy, A. (2009). 'Computational systems biology of the cell cycle'. In: *Brief. Bioinform.* 10.4, pp. 424–434.

Cushing, J. M. (1998). *An Introduction to Structured Population Dynamics*. Society for Industrial and Applied Mathematics. DOI: 10.1137/1.9781611970005.

Desoer, Charles A. and M. Vidyasagar (1975). *Feedback systems: Input-output properties*. Academic Press.

Elowitz, M B, A J Levine, E D Siggia and P S Swain (2002). 'Stochastic gene expression in a single cell'. In: *Science* 297.5584, pp. 1183–1186.

Ermentrout, Bard (2002). *Simulating, Analyzing, and Animating Dynamical Systems*. Society for Industrial and Applied Mathematics.

Evans, L.C. and American Mathematical Society (1998). *Partial Differential Equations*. Graduate studies in mathematics. American Mathematical Society.

Feillet, Celine, Gijsbertus T. J. van der Horst, Francis Levi, David A. Rand and Franck Delaunay (2015). 'Coupling between the Circadian Clock and Cell Cycle Oscillators: Implication for Healthy Cells and Malignant Growth'. In: *Frontiers in Neurology* 6.5, pp. 1–7. DOI: 10.3389/fneur.2015.00096.

Feller, William (1954). 'Diffusion processes in one dimension'. In: *Trans. Amer. Math. Soc.* 77, pp. 1–31.

Fischer, David S. et al. (2019). 'Inferring population dynamics from single-cell RNA-sequencing time series data'. In: *Nature Biotechnology* 37.4, pp. 461–468. DOI: 10.1038/s41587-019-0088-0.

Foerster, Heinz von (1959). 'Some remarks on Changing Populations'. In: *Kinet. Cell. Prolif.* Ed. by J. F. Stohlman. New York: Grune and Stratton, pp. 382–407.

Freyer, James P. and Robert M. Sutherland (1986). 'Regulation of Growth Saturation and Development of Necrosis in EMT6/Ro Multicellular Spheroids by the Glucose and Oxygen Supply'. In: *Cancer Research* 46.7, pp. 3504–3512.

Fröhlich, Fabian, Barbara Kaltenbacher, Fabian J Theis and Jan Hasenauer (2017). 'Scalable Parameter Estimation for Genome-Scale Biochemical Reaction Networks'. In: *PLOS Computational Biology* 13.1. Ed. by Jorg Stelling, e1005331. DOI: 10.1371/journal.pcbi.1005331.

Gabriel, Pierre, Shawn P. Garbett, Vito Quaranta, Darren R. Tyson and Glenn F. Webb (2012). 'The contribution of age structure to cell population responses to targeted therapeutics'. In: *Journal of Theoretical Biology* 311, pp. 19–27.

Gérard, C. and Albert Goldbeter (2011). 'A skeleton model for the network of cyclin-dependent kinases driving the mammalian cell cycle'. In: *Interface Focus* 1.1, pp. 24–35.

Gérard, Claude, Didier Gonze and Albert Goldbeter (2012). 'Effect of positive feedback loops on the robustness of oscillations in the network of cyclin-dependent kinases driving the mammalian cell cycle'. In: *FEBS Journal* 279.18, pp. 3411–3431.

Giesen, Charlotte et al. (2014). 'Highly multiplexed imaging of tumor tissues with subcellular resolution by mass cytometry'. In: *Nature Methods* 11.4, pp. 417–422. DOI: 10.1038/nmeth.2869.

Glass, Leon (2001). 'Synchronization and rhythmic processes in physiology'. In: *Nature* 410.6825, pp. 277–284. DOI: 10.1038/35065745.

Green, N.J.B. (1988). 'On the simulation of diffusion processes close to boundaries'. In: *Molecular Physics* 65.6, pp. 1399–1408. DOI: 10.1080/00268978800101871.

Grenander, U. and G. Szegő (1958). *Toeplitz forms and their applications*. Univ of California Press, 245 p.

Guckenheimer, J. (1975). 'Isochrons and phaseless sets'. In: *J. Math. Biol.* 1.3, pp. 259–273.

Gut, Gabriele, Michelle D Tadmor, Dana Pe'er, Lucas Pelkmans and Prisca Liberali (2015). 'Trajectories of cell-cycle progression from fixed cell populations'. In: *Nature Methods* 12.10, pp. 951–954.

Gyllenberg, M. and G. F. Webb (1990). 'A nonlinear structured population model of tumor growth with quiescence'. In: *Journal of Mathematical Biology* 28.6, pp. 671–694.

Haario, Heikki, Marko Laine, Antonietta Mira and Eero Saksman (2006). 'DRAM: Efficient adaptive MCMC'. In: *Statistics and Computing* 16.4, pp. 339–354. DOI: 10.1007/s11222-006-9438-0.

Haghverdi, Laleh, Florian Buettner and Fabian J. Theis (2014). 'Diffusion maps for high-dimensional single-cell analysis of differentiation data'. In: *Bioinformatics* 31.18, pp. 2989–2998. DOI: 10.1093/bioinformatics/btv325.

Haghverdi, Laleh, Maren Büttner, F Alexander Wolf, Florian Buettner and Fabian J Theis (2016). 'Diffusion pseudotime robustly reconstructs lineage branching'. In: *Nature Methods* 13.10, pp. 845–848. DOI: 10.1038/nmeth.3971.

Hanahan, Douglas and Robert A Weinberg (2011). 'Hallmarks of cancer: the next generation.' In: *Cell* 144.5, pp. 646–74.

Hansen, Nikolaus (2006). 'The CMA Evolution Strategy: A Comparing Review'. In: *Towards a New Evolutionary Computation: Advances in the Estimation of Distribution Algorithms*. Ed. by Jose A. Lozano, Pedro Larrañaga, Iñaki Inza and Endika Bengoetxea. Berlin, Heidelberg: Springer Berlin Heidelberg, pp. 75–102. DOI: 10.1007/3-540-32494-1_4.

Hodgkin, A. L. and A. F. Huxley (1952). 'A quantitative description of membrane current and its application to conduction and excitation in nerve'. In: *J. Physiol.* 117.4, pp. 500–544. DOI: 10.1113/jphysiol.1952.sp004764.

Hofmann, Lorenz, Martin Ebert, Peter Alexander Tass and Christian Hauptmann (2011). 'Modified Pulse Shapes for Effective Neural Stimulation'. In: *Frontiers in Neuroengineering* 4.9, pp. 1–10. DOI: 10.3389/fneng.2011.00009.

Holland, Christian H. et al. (2020). 'Robustness and applicability of transcription factor and pathway analysis tools on single-cell RNA-seq data'. In: *Genome Biology* 21.1, pp. 1–19. DOI: 10.1186/s13059-020-1949-z.

Hoppensteadt, Frank C. and Eugene M. Izhikevich (1997). *Weakly Connected Neural Networks*. Vol. 126. Applied Mathematical Sciences. Springer New York.

Hross, Sabrina and Jan Hasenauer (2016). 'Analysis of CFSE time-series data using division-, age- and label-structured population models'. In: *Bioinformatics* 32.15, pp. 2321–2329. DOI: 10.1093/bioinformatics/btw131.

Huh, Dann and Johan Paulsson (2011). 'Non-genetic heterogeneity from stochastic partitioning at cell division'. In: *Nature Genetics* 43.2, pp. 95–100. DOI: 10.1038/ng.729.

Imig, D., N. Pollak, T. Strecker, P. Scheurich, F Allgöwer and S. Waldherr (2015). 'An individual-based simulation framework for dynamic, heterogeneous cell populations during extrinsic stimulations'. In: *J. Coupled Syst. Multiscale Dyn.* 3.2, pp. 143–155.

Iversen, L. et al. (2014). 'Ras activation by SOS: Allosteric regulation by altered fluctuation dynamics'. In: *Science* 345.6192, pp. 50–54.

Izhikevich, Eugene M (2007). *Dynamical systems in neuroscience*. MIT press.

Jabs, Julia et al. (2017). 'Screening drug effects in patient-derived cancer cells links organoid responses to genome alterations'. In: *Molecular Systems Biology* 13.11, p. 955. DOI: 10.15252/msb.20177697.

Kafri, Ran, Jason Levy, Miriam B. Ginzberg, Seungeun Oh, Galit Lahav and Marc W. Kirschner (2013). 'Dynamics extracted from fixed cells reveal feedback linking cell growth to cell cycle'. en. In: *Nature* 494.7438, pp. 480–483. DOI: 10.1038/nature11897.

Kapuy, Orsolya, Enuo He, Sandra López-Avilés, Frank Uhlmann, John J. Tyson and Béla Novák (2009). 'System-level feedbacks control cell cycle progression'. In: *FEBS Letters* 583.24, pp. 3992–3998.

Khalil, Hassan K (2002). *Nonlinear systems; 3rd ed.* Upper Saddle River, NJ: Prentice-Hall.

Kiss, István Z., Mark Quigg, Shi-Hyung Calvin Chun, Hiroshi Kori and John L. Hudson (2008). 'Characterization of Synchronization in Interacting Groups of Oscillators: Application to Seizures'. In: *Biophysical Journal* 94.3, pp. 1121–1130. DOI: 10.1529/biophysj.107.113001.

Kitano, Hiroaki (2002). 'Computational systems biology'. In: *Nature* 420.6912, pp. 206–210. DOI: 10.1038/nature01254.

Klein, Allon M. et al. (2015). 'Droplet Barcoding for Single-Cell Transcriptomics Applied to Embryonic Stem Cells'. In: *Cell* 161.5, pp. 1187–1201. DOI: 10.1016/j.cell.2015.04.044.

Klipp, Edda, Wolfram Liebermeister, Christoph Wierling, Axel Kowald, Hans Lehrach and Ralf Herwig (2009). *Systems Biology: A Textbook*. Weinheim: Wiley-VCH.

Krasovskii, Nicolai Nikolaevich (1963). *Stability of motion*. Vol. 2. Stanford university press Stanford.

Kuramoto, Yoshiki (1975). 'Self-entrainment of a population of coupled non-linear oscillators'. In: *Int. Symp. Math. Probl. Theor. Phys.* Vol. 39. Berlin/Heidelberg: Springer-Verlag, pp. 420–422. DOI: 10.1007/BFb0013365.

Kuramoto, Yoshiki (1984). *Chemical Oscillations, Waves, and Turbulence*. Vol. 19. Springer Series in Synergetics. Springer Berlin Heidelberg.

La Manno, Gioele et al. (2018). 'RNA velocity of single cells'. In: *Nature* 560.7719, pp. 494–498. DOI: 10.1038/s41586-018-0414-6.

La Salle, Joseph P (1966). 'An invariance principle in the theory of stability'. In:

LaBarbera, Daniel V, Brian G Reid and Byong Hoon Yoo (2012). 'The multicellular tumor spheroid model for high-throughput cancer drug discovery'. In: *Expert Opinion on Drug Discovery* 7.9, pp. 819–830. DOI: 10.1517/17460441.2012.708334.

Levskaya, A, O D Weiner, W A Lim and C A Voigt (2010). 'Spatiotemporal control of cell signalling using a light-switchable protein interaction'. In: *Nature* 461.7266, pp. 1–5.

Li, L. and D.V. LaBarbera (2017). '3D High-Content Screening of Organoids for Drug Discovery'. In: *Comprehensive Medicinal Chemistry III* December 2016, pp. 388–415. DOI: 10.1016/B978-0-12-409547-2.12329-7.

Li, Jr-Shin, Isuru Dasanayake and Justin Ruths (2013). 'Control and synchronization of neuron ensembles'. In: *IEEE Transactions on Automatic Control* 58.8, pp. 1919–1930. DOI: 10.1109/TAC.2013.2250112.

Lin, Ruei-Zhen and Hwan-You Chang (2008). 'Recent advances in three-dimensional multicellular spheroid culture for biomedical research'. In: *Biotechnology Journal* 3.9-10, pp. 1172–1184. DOI: 10.1002/biot.200700228.

Liu, Pengda et al. (2014). 'Cell-cycle-regulated activation of Akt kinase by phosphorylation at its carboxyl terminus'. In: *Nature* 508.7497, pp. 541–545. DOI: 10.1038/nature13079.

Liu, Yin et al. (2017). 'Transcriptional landscape of the human cell cycle'. In: *Proceedings of the National Academy of Sciences* 114.13, pp. 3473–3478. DOI: 10.1073/pnas.1617636114.

Luecken, Malte D and Fabian J Theis (2019). 'Current best practices in single-cell RNA-seq analysis: a tutorial'. In: *Molecular Systems Biology* 15.6, e8746. DOI: 10.15252/msb.20188746.

MacArthur, Ben D., Avi Maayan and Ihor R. Lemischka (2009). 'Systems biology of stem cell fate and cellular reprogramming'. In: *Nature Reviews Molecular Cell Biology* 10.10, pp. 672–681. DOI: 10.1038/nrm2766.

Malkin, I. G. (1949). *Methods of Poincare and Liapunov in theory of non-linear oscillations*. Moscow: Gostexizdat.

Malkin, I. G. (1956). *Some Problems in Nonlinear Oscillation Theory*. Moscow: Gostexizdat.

Matchen, T. and J. Moehlis (2017). 'Real-time stabilization of neurons into clusters'. In: *2017 American Control Conference (ACC)*, pp. 2805–2810. DOI: 10.23919/ACC.2017.7963376.

McKendrick, A. G. (1926). 'Applications of Mathematics to Medical Problems'. In: *Proc. Edinburgh Math. Soc.* 44, pp. 98–130.

Metz, J. A. J. and O. Diekmann, eds. (1986). *The Dynamics of Physiologically Structured Populations*. Vol. 68. Lecture Notes in Biomathematics. Berlin, Heidelberg: Springer Berlin Heidelberg.

Mirollo, Renato E and Steven H Strogatz (1990). 'Synchronization of Pulse-Coupled Biological Oscillators'. In: *SIAM J. Appl. Math.* 50.6, pp. 1645–1662.

Mirsky, H. P., A. C. Liu, D. K. Welsh, S. A. Kay and F. J. Doyle (2009). 'A model of the cell-autonomous mammalian circadian clock'. In: *Proc. Natl. Acad. Sci.* 106.27, pp. 11107–11112. DOI: 10.1073/pnas.0904837106.

Monga, B., G. Froyland and J. Moehlis (2018). 'Synchronizing and Desynchronizing Neural Populations through Phase Distribution Control'. In: *2018 Annual American Control Conference (ACC)*, pp. 2808–2813. DOI: 10.23919/ACC.2018.8431114.

Monga, Bharat and Jeff Moehlis (2019). 'Phase distribution control of a population of oscillators'. In: *Physica D: Nonlinear Phenomena* 398, pp. 115–129. DOI: https://doi.org/10.1016/j.physd.2019.06.001.

Montenbruck, Jan Maximilian, Mathias Bürger and Frank Allgöwer (2015). 'Practical synchronization with diffusive couplings'. In: *Automatica* 53, pp. 235–243.

Morgan, David O (2007). *The cell cycle: Principles of control*. London: New Science Press.

Munsky, Brian, Brooke Trinh and Mustafa Khammash (2009). 'Listening to the noise: random fluctuations reveal gene network parameters'. In: *Molecular Systems Biology* 5.318, pp. 1–7.

Norbury, C. and P. Nurse (1992). 'Animal Cell Cycles and Their Control'. In: *Annual Review of Biochemistry* 61.1, pp. 441–468.

Pampaloni, Francesco, Emmanuel G. Reynaud and Ernst H K Stelzer (2007). 'The third dimension bridges the gap between cell culture and live tissue'. In: *Nature Reviews Molecular Cell Biology* 8.10, pp. 839–845. DOI: 10.1038/nrm2236.

Papili Gao, Nan, S M Minhaz Ud-Dean, Olivier Gandrillon and Rudiyanto Gunawan (2017). 'SINCERITIES: inferring gene regulatory networks from time-stamped single cell transcriptional expression profiles'. In: *Bioinformatics* 34.2, pp. 258–266. DOI: 10.1093/bioinformatics/btx575.

Papili Gao, Nan, Thomas Hartmann, Tao Fang and Rudiyanto Gunawan (2019). 'CALISTA: Clustering And Lineage Inference in Single-Cell Transcriptional Analysis'. In: *bioRxiv*. DOI: 10.1101/257550.

Perthame, Benoît and Jorge P Zubelli (2007). 'On the inverse problem for a size-structured population model'. In: *Inverse Problems* 23.3, pp. 1037–1052.

Polya, George and Norbert Wiener (1942). 'On the Oscillation of the Derivatives of A Periodic Function'. In: *Trans. Am. Math. Soc.* 52.2, pp. 249–256.

Pomerening, Joseph R., Eduardo D. Sontag and James E. Ferrell (2003). 'Building a cell cycle oscillator: hysteresis and bistability in the activation of Cdc2'. In: *Nature Cell Biology* 5.4, pp. 346–351.

Powell, E. O. (1956). 'Growth Rate and Generation Time of Bacteria, with Special Reference to Continuous Culture'. In: *Journal of General Microbiology* 15.3, pp. 492–511. DOI: 10.1099/00221287-15-3-492.

Qiu, Xiaojie, Andrew Hill, Jonathan Packer, Dejun Lin, Yi-An Ma and Cole Trapnell (2017). 'Single-cell mRNA quantification and differential analysis with Census'. In: *Nature Methods* 14.3, pp. 309–315. DOI: 10.1038/nmeth.4150.

Raue, Andreas et al. (2013). 'Lessons Learned from Quantitative Dynamical Modeling in Systems Biology'. In: *PLoS ONE* 8.9. Ed. by Enrique Hernandez-Lemus, e74335. DOI: 10.1371/journal.pone.0074335.

Redner, Sidney (2001). *A Guide to First-Passage Processes:* Cambridge University Press. DOI: 10.1017/CBO9780511606014.

Saelens, Wouter, Robrecht Cannoodt, Helena Todorov and Yvan Saeys (2019). 'A comparison of single-cell trajectory inference methods'. In: *Nature Biotechnology* 37, pp. 547–554. DOI: 10.1038/s41587-019-0071-9.

Saez-Rodriguez, Julio and Nils Blüthgen (2020). 'Personalized signaling models for personalized treatments'. In: *Molecular Systems Biology* 16.1, pp. 2–4. DOI: 10.15252/msb.20199042.

Sakaue-Sawano, Asako et al. (2008). 'Visualizing Spatiotemporal Dynamics of Multicellular Cell-Cycle Progression'. In: *Cell* 132.3, pp. 487–498. DOI: 10.1016/j.cell.2007.12.033.

Sandler, Oded, Sivan Pearl Mizrahi, Noga Weiss, Oded Agam, Itamar Simon and Nathalie Q. Balaban (2015). 'Lineage correlations of single cell division time as a probe of cell-cycle dynamics'. In: *Nature* 519.7544, pp. 468–471.

Santos, Silvia D M, Roy Wollman, Tobias Meyer and James E. Ferrell (2012). 'Spatial positive feedback at the onset of mitosis'. In: *Cell* 149.7, pp. 1500–1513.

Satija, Rahul, Jeffrey A Farrell, David Gennert, Alexander F Schier and Aviv Regev (2015). 'Spatial reconstruction of single-cell gene expression data'. In: *Nature Biotechnology* 33.5, pp. 495–502. DOI: 10.1038/nbt.3192.

Scardovi, Luca, Murat Arcak and Eduardo D Sontag (2010). 'Synchronization of Interconnected Systems With Applications to Biochemical Networks: An Input-Output Approach'. In: *IEEE Trans. Automat. Contr.* 55.6, pp. 1367–1379.

Setty, Manu et al. (2016). 'Wishbone identifies bifurcating developmental trajectories from single-cell data'. In: *Nature Biotechnology* 34.6, pp. 637–645. DOI: 10.1038/nbt.3569.

Sharpe, F.R. and .J. Lotka (1911). 'L. A problem in age-distribution'. In: *Philosophical Magazine Series 6* 21.124, pp. 435–438.

Sigal, Alex et al. (2006). 'Dynamic proteomics in individual human cells uncovers widespread cell-cycle dependence of nuclear proteins.' In: *Nature methods* 3.7, pp. 525–531.

Silverman, B.W. (1986). *Density Estimation for Statistics and Data Analysis*. Chapman & Hall/CRC Monographs on Statistics & Applied Probability. Taylor & Francis.

Singer, A., Z. Schuss, A. Osipov and D. Holcman (2008). 'Partially Reflected Diffusion'. In: *SIAM Journal on Applied Mathematics* 68.3, pp. 844–868. DOI: 10.1137/060663258.

Skorokhod, A. V. (1961). 'Stochastic Equations for Diffusion Processes in a Bounded Region'. In: *Theory of Probability & Its Applications* 6.3, pp. 264–274.

Snijder, Berend, Raphael Sacher, Pauli Rämö, Eva-Maria Damm, Prisca Liberali and Lucas Pelkmans (2009). 'Population context determines cell-to-cell variability in endocytosis and virus infection'. In: *Nature* 461.7263, pp. 520–523.

St Hilaire, Melissa A., Joshua J. Gooley, Sat Bir S. Khalsa, Richard E. Kronauer, Charles A. Czeisler and Steven W. Lockley (2012). 'Human phase response curve to a 1 h pulse of bright white light'. In: *The Journal of Physiology* 590.13, pp. 3035–3045. DOI: 10.1113/jphysiol.2012.227892.

Stiefel, Klaus M., Boris S. Gutkin and Terrence J. Sejnowski (2008). 'Cholinergic Neuromodulation Changes Phase Response Curve Shape and Type in Cortical Pyramidal Neurons'. In: *PLOS ONE* 3.12, pp. 1–7. DOI: 10.1371/journal.pone.0003947.

Strell, Carina, Markus M. Hilscher, Navya Laxman, Jessica Svedlund, Chenglin Wu, Chika Yokota and Mats Nilsson (2019). 'Placing RNA in context and space – methods for spatially resolved transcriptomics'. In: *FEBS Journal* 286.8, pp. 1468–1481. DOI: 10.1111/febs.14435.

Swain, Peter S., Michael B. Elowitz and Eric D. Siggia (2002). 'Intrinsic and extrinsic contributions to stochasticity in gene expression'. In: *Proceedings of the National Academy of Sciences* 99.20, pp. 12795–12800.

Szalai, Bence, Vigneshwari Subramanian, Christian H. Holland, Róbert Alföldi, László G. Puskás and Julio Saez-Rodriguez (2019). 'Signatures of cell death and proliferation in perturbation transcriptomics data - from confounding factor to effective prediction'. In: *Nucleic Acids Research* 47.19, pp. 10010–10026. DOI: 10.1093/nar/gkz805.

Tanay, Amos and Aviv Regev (2017). 'Scaling single-cell genomics from phenomenology to mechanism'. In: *Nature* 541.7637, pp. 331–338. DOI: 10.1038/nature21350.

Theis, Fabian J., Florian Buettner and Laleh Haghverdi (2015). 'Diffusion maps for high-dimensional single-cell analysis of differentiation data'. In: *Bioinformatics* 31.18, pp. 2989–2998. DOI: 10.1093/bioinformatics/btv325.

Trapnell, Cole et al. (2014). 'The dynamics and regulators of cell fate decisions are revealed by pseudotemporal ordering of single cells'. In: *Nature Biotechnology* 32.4, pp. 381–386. DOI: 10.1038/nbt.2859.

Tritschler, Sophie, Maren Büttner, David S. Fischer, Marius Lange, Volker Bergen, Heiko Lickert and Fabian J. Theis (2019). 'Concepts and limitations for learning developmental trajectories from single cell genomics'. In: *Development* 146.12. DOI: 10.1242/dev.170506.

Vörsmann, H., F. Groeber, H. Walles, S. Busch, S. Beissert, H. Walczak and D. Kulms (2013). 'Development of a human three-dimensional organotypic skin-melanoma spheroid model for in vitro drug testing'. In: *Cell Death & Disease* 4.7, e719. DOI: 10.1038/cddis.2013.249.

Vosko, Andrew M, Alon Avidan and Christopher Colwell (2010). 'Jet lag syndrome: circadian organization, pathophysiology, and management strategies'. In: *Nature and Science of Sleep* 2, p. 187. DOI: 10.2147/NSS.S6683.

Weinreb, Caleb, Samuel Wolock, Betsabeh K. Tusi, Merav Socolovsky and Allon M. Klein (2018). 'Fundamental limits on dynamic inference from single-cell snapshots'. In: *Proceedings of the National Academy of Sciences* 115.10, E2467–E2476. DOI: 10.1073/pnas.1714723115.

Wheeler, Richard John (2015). 'Analyzing the dynamics of cell cycle processes from fixed samples through ergodic principles.' In: *Molecular Biology of the Cell* 26.22, pp. 3898–903. DOI: 10.1091/mbc.E15-03-0151.

Wiener, Norbert (1938). 'The Homogeneous Chaos'. In: *American Journal of Mathematics* 60.4, pp. 897–936.

Wijst, Monique GP van der et al. (2020). 'The single-cell eQTLGen consortium'. In: *eLife* 9. DOI: 10.7554/eLife.52155.

Wilson, Dan and Jeff Moehlis (2014). 'Optimal Chaotic Desynchronization for Neural Populations'. In: *SIAM J. Appl. Dyn. Syst.* 13.1, pp. 276–305.

Wilson, Dan and Jeff Moehlis (2015). 'Determining individual phase response curves from aggregate population data'. In: *Phys. Rev. E* 92.2, p. 022902. DOI: 10.1103/PhysRevE.92.022902.

Wilson, Dan and Jeff Moehlis (2016). 'Isostable reduction with applications to time-dependent partial differential equations'. In: *Phys. Rev. E* 94.1, pp. 1–14. DOI: 10.1103/PhysRevE.94.012211.

Winfree, A. T. (1974). 'Patterns of phase compromise in biological cycles'. In: *J. Math. Biol.* 1.1, pp. 73–93.

Winfree, Arthur T (1967). 'Biological rhythms and the behavior of populations of coupled oscillators'. In: *J. Theor. Biol.* 16.1, pp. 15–42. DOI: 10.1016/0022-5193(67)90051-3.

Winfree, Arthur T. (1986). *Timing of Biological Clocks*. Henry Holt and Company, p. 199.

Winfree, Arthur T. (2001). *The Geometry of Biological Time*. Vol. 12. Interdisciplinary Applied Mathematics. New York, NY: Springer New York. DOI: 10.1007/978-1-4757-3484-3.

Xeros, N. (1962). 'Deoxyriboside Control and Synchronization of Mitosis'. In: *Nature* 194.4829, pp. 682–683.

Zeng, Shen and Frank Allgöwer (2016). 'A moment-based approach to ensemble controllability of linear systems'. In: *Systems & Control Letters* 98, pp. 49–56.

Zeng, Shen, Steffen Waldherr, Christian Ebenbauer and Frank Allgower (2016). 'Ensemble Observability of Linear Systems'. In: *IEEE Trans. Automat. Contr.* 61.6, pp. 1452–1465.

Zhivotovsky, B and S Orrenius (2010). 'Cell cycle and cell death in disease: past, present and future'. In: *J. Intern. Med.* 268.5, pp. 395–409.

Zlotnik, Anatoly and Jr-Shin Li (2014). 'Optimal Subharmonic Entrainment of Weakly Forced Nonlinear Oscillators'. In: *SIAM Journal on Applied Dynamical Systems* 13.4, pp. 1654–1693. DOI: 10.1137/140952211.

Publications of the Author

Colbrook, Matthew J., Zdravko I. Botev, Karsten Kuritz and Shev MacNamara (2020). 'Kernel density estimation with linked boundary conditions'. In: *Studies in Applied Mathematics* 145.3, pp. 357–396. DOI: 10.1111/sapm.12322.

Kuritz, Karsten (2020). *MAPiT: measure-preserving MAP of pseudotime into true Time*. Version v1.0. DOI: 10.5281/zenodo.3630379.

Kuritz, Karsten, Alain R Bonny, João Pedro Fonseca and Frank Allgöwer (2020a). 'PDE-constrained optimization for estimating population dynamics over cell cycle from static single cell measurements'. In: *bioRxiv*. DOI: 10.1101/2020.03.30.015909.

Kuritz, Karsten, Daniela Stöhr, Daniela Simone Maichl, Nadine Pollak, Markus Rehm and Frank Allgöwer (2020b). 'Reconstructing temporal and spatial dynamics from single-cell pseudotime using prior knowledge of real scale cell densities'. In: *Scientific Reports* 10.1, p. 3619. DOI: 10.1038/s41598-020-60400-z.

Kuritz, Karsten, Shen Zeng and Frank Allgöwer (2019). 'Ensemble Controllability of Cellular Oscillators'. In: *IEEE Control Systems Letters* 3.2, pp. 296–301. DOI: 10.1109/lcsys.2018.2870967.

Kuritz, Karsten, Wolfgang Halter and Frank Allgöwer (2018a). 'Passivity-based ensemble control for cell cycle synchronization'. In: *Emerg. Appl. Control Syst. Theory*. Ed. by Roberto Tempo, Stephen Yurkovich and Pradeep Misra. 1st ed. Springer International Publishing. DOI: 10.1007/978-3-319-67068-3_1.

Kuritz, Karsten, Dirke Imig, Michael Dyck and Frank Allgöwer (2018b). 'Ensemble control for cell cycle synchronization of heterogeneous cell populations'. In: *IFAC-PapersOnLine* 51.19. 7th Conference on Foundation of Systems Biology in Engineering FOSBE 2018, pp. 44–47. DOI: 10.1016/j.ifacol.2018.09.034.

Kuritz, Karsten, Daniela Stöhr, Nadine Pollak and Frank Allgöwer (2017). 'On the relationship between cell cycle analysis with ergodic principles and age-structured cell population models'. In: *Journal of Theoretical Biology* 414, pp. 91–102. DOI: 10.1016/j.jtbi.2016.11.024.